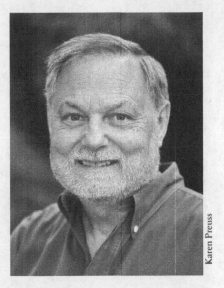

Karen Preuss

DUANE ELGIN is a visionary author, speaker, evolutionary activist, and educator. The author of *Voluntary Simplicity*, *Promise Ahead*, *Awakening Earth*, and *The Living Universe*, he is a former senior social scientist at SRI International, where he coauthored numerous studies on the long-range future for the Environmental Protection Agency, the President's Science Adviser, and the National Science Foundation. Prior to working at SRI, Elgin was a senior staff member at the National Commission on Population Growth and the American Future. He holds an MBA from the Wharton School of Business and an MA in economic history from the University of Pennsylvania. He lives in the San Francisco Bay Area.

VOLUNTARY SIMPLICITY

also by **DUANE ELGIN**   *The Living Universe*
Promise Ahead
Awakening Earth

HARPER

NEW YORK ● LONDON ● TORONTO ● SYDNEY

VOLUNTARY
SIMPLICITY

TOWARD A WAY OF LIFE

THAT IS OUTWARDLY SIMPLE,

INWARDLY RICH

Second Revised Edition

Duane Elgin

HARPER

First edition printed in 1981.

A revised paperback edition of this book was published in 1993 by Harper-Collins Publishers.

VOLUNTARY SIMPLICITY (Second Revised Edition). Copyright © 2010 by Duane Elgin. All rights reserved. Printed in the United States of America. No part of this book may be used or reproduced in any manner whatsoever without written permission except in the case of brief quotations embodied in critical articles and reviews. For information, address HarperCollins Publishers, 195 Broadway, New York, NY 10007..

HarperCollins books may be purchased for educational, business, or sales promotional use. For information please e-mail Special Markets Department, at SPsales@harpercollins.com..

SECOND REVISED EDITION

Designed by Janet M. Evans

Library of Congress Cataloging-in-Publication Data is available upon request.

ISBN 978-0-06-177926-8 (second revised edition)

HB 09.22.2021

Contents

Acknowledgments

There are innumerable persons who have helped me along the way—either in the learning that led to the writing of this book or in the writing process itself. I want to acknowledge particularly: my father for his integrity, generosity, and example of patient craftsmanship; my mother for her curiosity, compassion, and zest for life; Daniel Berrigan for teaching me that the way of love does not turn from life; Donald Michael for his enthusiasm for learning with an open mind and open heart; Arnold Mitchell for his collaboration in early research on simpler living as well as his helpful critiques of the early drafts of this book; Annie Folger for her loving support; Ram Dass for his patient reading of early versions of this book and for writing the insightful foreword; Frances Vaughan for her encouragement and enthusiasm for this work; Roger Walsh for his friendship, support, and discerning reading of numerous versions of this manuscript; and Ronda Davé for her skilled computer analysis of the survey results. A huge thanks to all of the people who responded to the grassroots survey for bringing a richly human dimension to this work. Finally, I want to acknowledge the assistance of other persons who, in various ways, have helped this work along its path to completion: Barry Bartlett, John Brockman, Joe Dominguez, Andrew Dutter, Coleen LeDrew Elgin, Sally Furgeson, Paula Hendrick, Bertil Ilhage, Kathryn Reeder, Vicki Robin, Mary Schoonmaker, Peter Schwartz, Peter Teige, Mary Thomas, John White, and Monica Wood.

Foreword

to the Second Revised Edition

EDGAR MITCHELL

In 1971, during my return from the moon aboard the Apollo 14 command module, I looked out the window at the space, stars, and planet from which I'd come and suddenly experienced the universe as intelligent, loving, and harmonious. I also realized that the Earth is a gem in the cosmos, a place to revere and care for. We are here not by accident, but on a journey of awakening that is as magnificent as the universe that holds us. We have a profound responsibility to care for the Earth, which is our craft on a voyage of both outer and inner discovery. The visionary scientist-philosopher Buckminster Fuller, at the beginning of space flight, pointed out: "We are the crew of space-ship Earth, but we are a crew in mutiny. How can you run a spacecraft with a mutinous crew?" Duane Elgin's book helps to quell this mutiny against nature, as he compassionately confronts the difficult challenge of making a transition to a new world in which we are living in balance with the Earth.

I met Duane in 1973 when he was working as a senior social scientist doing "futures studies" at the California think tank SRI

International. His primary work was to look ahead a generation or more and explore a range of alternative futures for government agencies such as the National Science Foundation and the President's Science Adviser. Duane was also working at SRI as one of four principal subjects in experiments being funded by NASA to explore our intuitive or psychic potentials. This was an area of keen interest to me, as I had conducted such experiments in space while on my lunar mission and I was in the process of founding the Institute of Noetic Sciences (www.noetic.org) to explore the nature of consciousness. Duane and I discovered mutual interests not only in exploring untapped human potentials but also in our concern for humanity's future. For more than thirty years, he has explored the frontiers of human knowledge to understand the nature of the universe and the human journey within it. Our paths have crossed a number of times, and I have grown to appreciate him and his work as a scholar, visionary, and activist. Duane's love for the Earth and the human journey is apparent.

I believe this book is important because we are overconsuming the planet and we are not sustaining our spacecraft Earth for future generations. Within a generation we can devastate the ecological foundations of the planet, or we can learn to live and consume within the boundaries of what the Earth can support. The days have now passed when we can ignore the impact of our individual decisions and lifestyles upon the collective. *Personal responsibility for the greater good must become the mark of an informed and conscious people.* It is vital to our future as a human family that we each look beyond our personal lifestyles and consider the well-being of the whole. Unrestricted consumption and growth in all areas of life must be reexamined and subjected to critical thought. We must act quickly to bring our human desire for material abundance under control. Otherwise, we risk wounding the biosphere so profoundly that the foundation for a healthy and robust human future is forever diminished. I pray that our descendants will forgive us these errors as we struggle to bring our Earth back into balance.

Duane's book offers a realistic pathway for a hopeful future. To me, a world of sophisticated simplicity sounds interesting, collaborative, artistic, and creative. Instead of regress, Duane is calling for a new kind of progress that integrates the inner and outer aspects of our lives. This is an important call for Earth-friendly ways of living that are grounded in practical actions by caring individuals who understand the critical challenges facing humanity. Duane calls on us to take personal responsibility for the well-being of the Earth and for cocreating a future of soaring promise. I hope this book becomes a catalyst for the vitally important conversation that our world must have if we are to voluntarily bring our personal and collective lives into harmonious balance with the Earth.

EDGAR MITCHELL, Sc.D.,
Ph.D.; astronaut, Apollo 14;
founder and chairman,
Institute of Noetic Sciences
2009

Foreword

to the First Edition (1981)

RAM DASS

As one who has spent a great deal of time in the East, I've had the chance to view intimately a way of life that, in its simplicity, is very different from the style of living to which we in the West are accustomed. Even as I write these words, I look out over a gentle valley in the Kumoan Hills at the base of the Himalayas. A river flows through the valley, forming now and again man-made tributaries that irrigate the fertile fields. These fields surround the fifty or so thatched or tin-roofed houses and extend in increasingly narrow terraces up the surrounding hillsides.

In several of these fields I watch village men standing on their wooden plows goading on their slow-moving water buffalo, who pull the plows, provide the men's families with milk, and help carry their burdens. And amid the green of the hills, in brightly colored saris and nose rings, women cut the high grasses to feed the buffalo and gather the firewood that, along with the dried dung from the buffalo, will provide the fire to cook the grains harvested from the fields and to warm the houses against the winter cold and dry them during the monsoons. A huge hay-stack passes along the path, seemingly self-propelled, in that the woman on whose head it rests is entirely lost from view.

At a point along the stream there is laughter and talk and the continuous slapping of wet cloth against the rock as the family laundry gets done. And everywhere there are children and dogs, each contributing his or her sound to the voice of the village.

Everywhere there is color: red chili peppers drying on the roofs and saris drying by the river, small green and yellow and blue birds darting among the fruit trees, butterflies and bees tasting their way from one brightly colored flower to another.

I have walked for some five miles from the nearest town to reach this valley. The footpath I have taken is the only means of exit from this village. Along the way I meet farmers carrying squash, burros bearing firewood or supplies, women with brass pots on their heads, schoolchildren, young men dressed in "city clothes." In all of these people I find a quiet, shy dignity, a sense of belonging, a depth of connectedness to these ancient hills.

It all moves as if in slow motion. Time is measured by the sun, the seasons, and the generations. A conch shell sounds from a tiny temple, which houses a deity worshiped in these hills. The stories of this and other deities are recited and sung, and they are honored by flowers and festivals and fasts. They provide a context—vast in its scale of aeons of time, rich with teachings of reincarnation and the morality inherent in the inevitable workings of karma. And it is this context that gives vertical meaning to these villagers' lives, with their endless repetition of cycles of birth and death.

This pastoral vision of simplicity has much appeal to those of us in the West, for whom life can be full of confusion, distraction, and complexity. In the rush of modern industrial society and in the attempt to maintain our image as successful persons, we feel that we have lost touch with a deeper, more profound part of our being. But we feel that we have little time, energy, or cultural support to pursue areas of life that we know are important. We long for a simpler way of life that allows us to restore some balance to our lives.

Is the vision of simple living provided by this village in the East the answer? Is this an example of a primitive simplicity of the past or of an enlightened simplicity of the future?

Gradually I have come to sense that this is not the kind of simplicity that the future holds. For despite its ancient character, the simplicity of the village is still in its "infancy."

Occasionally people show me their new babies and ask me if that peaceful innocence is not just like that of the Buddha. Probably not, I tell them, for within that baby reside all the latent seeds of worldly desire, just waiting to sprout as the opportunity arises. On the other hand, the expression on the face of the Buddha, who had seen through the impermanence and suffering associated with such desires, reflects the invulnerability of true freedom.

So it is with this village. Its ecological and peaceful way of living is unconsciously won and thus is vulnerable to the winds of change that fan the latent desires of its people. Even now there is a familiar but jarring note in this sylvan village scene. The sounds of static and that impersonal professional voice of another civilization—the radio announcer—cut through the harmony of sounds as a young man of the village holding a portable radio to his ear comes around a bend. On his arm there is a silver wristwatch, which sparkles in the sun. He looks at me proudly as he passes. And a wave of understanding passes through me. Just behind that radio and wristwatch comes an army of desires that for centuries have gone untested and untasted. As material growth and technological change activate these yearnings, they will transform the hearts, minds, work, and daily life of this village within a generation or two.

Gradually I see that the simplicity of the village has not been consciously chosen as much as it has been unconsciously derived as the product of centuries of unchanging custom and tradition. The East has yet to fully encounter the impact of technological change and material growth. When the East has encountered the latent desires within its people, and the cravings for material goods and social position begin to wear away at the fabric of traditional culture, then it can begin to choose its simplicity consciously. Then the simplicity of the East will be consciously won—voluntarily chosen.

Just as the East still has its time of transition to move through before finding a new place of dynamic equilibrium and conscious balance, so, too, does the West face its own unique time of transition. In the West there are many who have already begun the search for a more conscious balance, a simplicity of living that allows the integration of inner and outer, material and spiritual, masculine and feminine, personal and social, and all of the other polarities that now divide our lives. Since the decade of the 1960s, this search for integration has led a whole subculture to explore the learnings of the East. Yet, in attempting to partake deeply of the rich heritage of the East, many, myself included, have tended to diminish the value and relevance of our Western heritage.

My mind flashes back to 1968, when I returned from my first trip to India. Through the eyes of a renunciate I saw comfort and convenience, aesthetics and pleasure, as the sirens seducing me back into the sleep of unconsciousness. So I moved into a cabin in the woods behind my father's house. Each day I bathed out of a pail of cold water, huddled in blankets against the night chill, slept on a thin mat on the floor, and cooked the same food of lentils and rice.

Not fifty yards away, in my father's house, there were empty bedrooms, electricity, warm showers, television, and tasty and varied hot meals cooked each day. I would look at that house in the evenings with disdain, which I now suspect was born mainly out of longing. I was running as hard as I could away from Western values even as I was studying in depth the *Bhagavad-Gita*, which says that one must honor one's unique life predicament and not imitate another's. Though in time I surrendered slowly back into that sensual, material ease of the West, I did so with some sense of being a fallen angel, of compromising with a technological society running out of control. I did not know where a skillful balance was to be found. I had been fortunate enough to taste deeply the most generous fruits of two richly developed cultural traditions. Often I felt them churning and pulling within my being as I searched for balance along the vast spectrum of potentials represented by these two perspectives. I felt the dualism within my being. I could not reject the West and embrace the East—both lived within me.

Similarly, many among the counterculture of the 1960s withdrew from the emptiness of the industrial Disneyland of worldly delights (where even when you won, you lost) but did not halt their interior journey with an existential renunciation of the West. Many among a whole generation have turned within to the heart and have begun to move beyond intellectual alienation and despair to directly encounter the place where we are all connected, where we are all one. And from that place many have felt drawn to lifestyles in which their contact with their fellow human beings, with nature, and with God could be renewed.

A cycle of learning is being completed. The time of withdrawal is moving into a time of return. The exploration of new ways of living that support new ways of being is a movement that arises from the awakening of compassion—the dawning realization that the fate of the individual is intimately connected with the fate of the whole. It is a movement that arises from the recognition that our task is not only to be here, in the NOW; it is also to now be HERE. And where "here" is must include the fact that we are inhabitants of an aging industrial civilization that is in great need of the insight, perspective, and creativity that the journey to the "East" (to the interior) can bring upon return to the West.

The dualism inherent in our thinking process (which pits materialism against spiritualism, West against East) must be transcended if we are truly to be the inheritors of our evolutionary legacy and the children of a new age. From this perspective the historic Western preoccupation with the intellect and with material consumption need not be viewed as "wrong" or "bad" or necessarily leading to our destruction. Rather, the Western orientation in living may be viewed as a necessary part of an evolutionary stage out of which yet another birth of higher consciousness—as an amalgam of East and West—might evolve. The Industrial Revolution, then, is part of a larger revolution in living. The West has made its contribution by providing the material basis of life needed to support the widespread unfolding of consciousness. The contribution of the East is the provision of insight into the nature of the conscious unfolding of consciousness. East and West require the learnings of each other if both are to evolve further and realize the potentials that arise in their integration and balance.

Yet what does this mean in worldly terms? What are the practical, down-to-earth expressions of this integration of inner and outer, East and West, personal growth and social transformation? Are there ways of living that express this integrative intention? We must begin by acknowledging there are no "right answers" to these questions. The process of integration of East and West has only just begun to infuse the popular culture in the last few decades. We have only just begun to enter a new age of discovery. A vast new frontier beckons. The "answers" that we seek will be of our own making—they are still in the process of being discovered in our own lives.

And yet, for many of us, merely saying this is not enough. We want to see more clearly the larger pattern into which our lives can fit more skillfully. We also want to see more precisely how we might adapt our daily lives to fit harmoniously into this larger pattern of evolution.

What kind of person might assist in the delicate midwifery of revealing to us the nature of the worldly expression of an integration of inner and outer, East and West? Certainly one would be required to have a foot in both the Eastern and the Western perspectives. One must have cultivated in oneself a compassionate consciousness—a balance of head and heart. I think it requires people like E. F. Schumacher, the author of *Small Is Beautiful*, who, with compassion and penetrating insight, challenged many of the traditional assumptions of Western industrial societies. And now, with this book on voluntary simplicity, we meet another of these beings, Duane Elgin. People such as Duane are afraid neither to immerse themselves in the problems and potentials of an advanced industrial civilization nor to immerse themselves in the meditative journey. They find in this balancing process neither despair nor Pollyannaish optimism, but an enthusiasm for the dance. They can reveal the complex dynamics of the existing world situation with a wisdom that reflects self-respect, integrity, compassion, and subtle discriminations of the intellect. Such beings look to the future and see through the smoke to the sunlight.

RAM DASS,
formerly Dr. Richard Alpert
1980

How the Times Have Changed!

Introduction to the Second Revised Edition

There has been a seismic shift in public interest in simpler, more sustainable ways of living since this book was first published. This accelerating change is summarized beautifully in two introductions I was given to business audiences, separated by a span of nearly thirty years. In 1977, I gave a talk to an audience of business executives on an emerging way of life called "voluntary simplicity," and was introduced as a "Wharton MBA who had gone bad." Although intended with good-natured humor, it was clear that I was viewed as a renegade MBA, outside the business mainstream. Then in 2005, I was introduced to another audience of business executives, but this time as a "Wharton MBA who had gone green." On this occasion, I was viewed as a pioneering MBA on the cutting edge of a revolution in sustainability.

Going from "bad" to "green" in those two introductions summarizes a transformation that I have been watching for nearly four decades. More specifically, there are a half-dozen fundamental changes in how simplicity of living is now viewed by business, government, and the public.

First, the public conversation about simplicity is shifting from complacency to urgency. In the 1970s, there was little public concern about climate change, massive famines, energy and water shortages, and more. Although these loomed on the horizon, the majority of people were focused on the "good life" in the short run. More than thirty years later, these are no longer problems for the distant future; they represent a critical challenge to the human community *now.* The more closely we look, the more compelling is the evidence that the human family has exceeded the ability of the Earth to support humanity's current levels of consumption, let alone that projected for the future. There must be dramatic, global changes in our overall approach to living and consuming if we are to avoid a future of immense calamity. Simplicity of living, by whatever name, is moving from an easily dismissed lifestyle fad to an approach to living that is recognized as a vital ingredient for building a sustainable and meaningful future.

Second, as people's sense of urgency has grown, interest in sustainable ways of living has soared, and simplicity has moved from the margins of society to the mainstream. Simpler or greener approaches to living are becoming part of everyday life and culture. Television programs on themes such as organic gardening, healthy cooking, and solar living are growing in popularity. Magazines with green themes for living are sprouting everywhere. College courses in green building and environmental management are blossoming. There has been an explosion in Internet websites and blogs concerned with restoring the Earth to health and building a more just and sustainable economy for the world. Overall, the "center of social gravity" is shifting rapidly and simpler, greener ways of living are of growing interest and concern to the mainstream of many societies.

Third, public understanding of simplicity has evolved from fantasy stereotypes to realistic examples and archetypes. In the 1980s, it was common for the mass media to characterize simplicity as a "back to the land" movement that turned away from technological progress (an inaccurate stereotype, as the sustainability movement has generated a wave of technological innovation that is now recognized as vital to a green future). Several decades later, in response to growing economic and ecological crises, people are looking for resilient and practical approaches to living that are uniquely adapted to different settings. Pushed by necessity, a discerning social intelligence is emerging that looks beyond shallow stereotypes to a diverse garden of expressions that offer realistic models of change for diverse people and circumstances.

Fourth, simplicity has moved from being regarded as a path of regress to being seen as a path to a new kind of progress and social vitality. When I first began speaking about simplicity, it was often dismissed as a nostalgic desire to return to the past as an antidote to the impersonality of the fast pace of the city life. Simplicity was seen as turning back the clock—a lifeway of regress, not progress. Decades later, instead of a return to the past, simplicity is seen increasingly as vital for building a workable and meaningful future. To illustrate, where the traditional economic wisdom assumed that consuming less meant fewer jobs, a new economic wisdom says that consuming moderately, differently, and intelligently will produce both sustainable jobs and a healthy world for the long run.

Fifth, there has been a dramatic expansion in the scope of simplicity as it has moved from a personal issue to a consideration vital to our collective future. Despite the media interest in alternative lifestyles in the 1960s and 1970s, the vast majority of early pioneers of sustainable ways of living were individuals and families. Many of these people felt alone and out of step with the consumer culture. However, with energy shortages, climate change, economic breakdowns, and more, the scale of public concern and attention has grown from the personal to the national and planetary. Now we are seeing the rapid growth of interest in ecovillages, cohousing communities, transition towns, state-level initiatives, federal programs, and global agreements. Simplicity of living is no longer a personal issue; it is a theme and concern woven into our lives at every scale.

Sixth, over the decades, simplicity is increasingly being defined by what it is *for* (connecting with and caring for life) instead of what it is *against* (destructive consumerism). In the 1980s, simplicity was seen primarily as "downshifting," or pulling back from the rat race of consumer society. Several decades later, there is a growing recognition of simplicity as "upshifting"—or moving beyond the rat race to the human race. Increasingly, the mainstream media and society are recognizing how people's search for happiness is taking them beyond consumerism to a more balanced and integrated approach to living.

Overall, the world has changed dramatically since I wrote the revised edition of *Voluntary Simplicity* in the early 1990s. To respond, I've completely revised this book—more than half of it is new material. It is my hope this new edition will extend the promising wisdom and healing force of simplicity to our imperiled world, for on the other side of the fast-emerging planetary systems crisis is a future bright with promise.

DUANE ELGIN,
May 2009

VOLUNTARY SIMPLICITY

one

COOL LIFESTYLE FOR
A HOT PLANET

Simplicity is the ultimate sophistication.
—*Leonardo da Vinci*

HOW SHALL WE LIVE?

Time is up! Wake-up alarms are ringing around the world with news ranging from economic breakdowns and the end of cheap oil to climate disruption, crop failures, and famines. The time has arrived for making dramatic changes in how we live. If we act swiftly and voluntarily, we can transform catastrophe into opportunity. Small steps alone will not be sufficient. We require a radical redesign of our urban environments that emphasizes localized economies, a fundamental overhaul in our energy systems, a more conscious democracy with the strength to make great changes, and much more.

As individuals, are we helpless in the face of such immense challenges? Do we feel there is little we can do? The reality is just the opposite—only changes in our individual lives can establish a resilient and strong foundation for a promising future.

The choice facing humanity is described in stark terms by Professor Jared Diamond in his prize-winning book *Collapse*. He writes that, one way or another, the world's environmental problems will get resolved within a generation. "The only question is whether they will become resolved in pleasant ways of our own choice, or in unpleasant ways not of our choice, such as warfare, genocide, starvation, disease epidemics, and collapses of societies."[1] Our choice as a species is straightforward and profound. We can awaken ourselves from the dream of limitless material growth and actively invent new ways to live within the material limits of the Earth. Or we can continue along our current path of denial and bargaining, using up precious decades, until we slam into an evolutionary wall and so profoundly wound the biosphere and human relations that it cripples humanity's evolutionary possibilities for millennia to come.

So one choice is to continue along our current path of increasingly unsustainable consumption, knowing that it leads to a future of ecological ruin, and the other is to confront the reality of unsustainable consumer societies, bring this taboo topic squarely into our public conversation, and search for realistic alternatives. This is an extremely difficult public conversation because it challenges the underlying paradigm of materialism and the self-image of nations who are identified as consumer societies. Nonetheless, the global dialogue regarding how we are all to live on this Earth has begun in earnest. To illustrate, world leaders in science, religion, and politics were calling, in 2008, for a new path to sustainability and ecological sanity. In politics the premier of China called upon rich countries "to shoulder the duty and responsibility to tackle climate change and alter their unsustainable lifestyle." In religion, the pope criticized developed nations for "squandering the world's resources in order to fuel an insatiable consumption." In science, the world's leading climatologist, James Hansen, warned that without a dramatic reduction in greenhouse gas emissions, we will create a dramatically different and far less hospitable planet for the people of the Earth.

After two hundred or more years of material growth, we are confronted with an unyielding question: *If the material consumption of a fraction of humanity is already harming the planet, is there an alternative path that enables all of humanity to live more lightly upon the Earth while experiencing a higher quality of life?* This question reaches deep into humanity's psyche and soul. Transforming our levels and patterns of consumption requires our looking directly into how we create our sense of identity and seek our happiness. Furthermore, because the ecological challenges we face are global in nature, so too must be our conversation concerning how we are to share the Earth with one another and the rest of life.

Despite the necessity for change, it is hard to believe we humans will turn away from the lure of materialism and growth until we collectively recognize that this path leads, as Professor Diamond warns, to "warfare, genocide, starvation, disease . . . and collapse." A turn also requires compelling visions of the future that act as beacons for our social imagination. We do not yet carry in our social imagination clear visions of the opportunities afforded by new forms of growth. Instead of visualizing how material limitation can draw out new levels of community and cooperation, many people see a life of greater "simplicity" as a path of sacrifice and regress.

Living within the limits that the Earth can sustain raises a fundamental question: Can we live more lightly on the material side of life while living with greater satisfaction and meaning on the nonmaterial side of life? In short, is simplicity a life of sacrifice?

SIMPLICITY IS NOT SACRIFICE

Simplicity that is *voluntary*—consciously chosen, deliberate, and intentional—supports a higher quality of life. Here are some reasons people consciously choose simplicity:

- Simplicity fosters a more harmonious relationship with the Earth—the land, air, and water.

- Simplicity promotes fairness and equity among the people of the Earth.

- Simplicity cuts through needless busyness, clutter, and complications.

- Simplicity enhances living with balance—inner and outer, work and family, family and community.

- Simplicity reveals the beauty and intelligence of nature's designs.

- Simplicity increases the resources available for future generations.

- Simplicity helps save animal and plant species from extinction.

* Simplicity responds to global shortages of oil, water, and other vital resources.

* Simplicity keeps our eyes on the prize of what matters most in our lives—the quality of our relationships with family, friends, community, nature, and cosmos.

* Simplicity yields lasting satisfactions that more than compensate for the fleeting pleasures of consumerism.

* Simplicity fosters the sanity of self-discovery and an integrated approach to life.

* Simplicity blossoms in community and connects us to the world with a sense of belonging and common purpose.

* Simplicity is a lighter lifestyle that fits elegantly into the real world of the twenty-first century.

Voluntary simplicity is not sacrifice:

* Sacrifice is a consumer lifestyle that is overstressed, overbusy, and overworked.

* Sacrifice is investing long hours doing work that is neither meaningful nor satisfying.

* Sacrifice is being apart from family and community to earn a living.

- Sacrifice is the stress of commuting long distances and coping with traffic.

- Sacrifice is the white noise of civilization blotting out the subtle sounds of nature.

- Sacrifice is hiding nature's beauty behind a jumble of billboard advertisements.

- Sacrifice is carrying more than two hundred toxic chemicals in our bodies, with consequences that will cascade for generations ahead.

- Sacrifice is the massive extinction of plants and animals and a dramatically impoverished biosphere.

- Sacrifice is being cut off from nature's wildness and wisdom.

- Sacrifice is global climate disruption, crop failure, famine, and forced migration.

- Sacrifice is the absence of feelings of neighborliness and community.

- Sacrifice is the lack of opportunity for soulful encounters with others.

- Sacrifice is feeling divided among the different parts of our lives and unsure how they work together in a coherent whole.

Contrary to media myths, consumerism offers lives of sacrifice while simplicity offers lives of opportunity. Simplicity creates the opportunity for greater fulfillment in work, meaningful connection with others, feelings of kinship with all life, and awe of a living universe. This is a rich way of life that offers a compelling alternative to the stress, busyness, and alienation of the modern era. However, the mainstream media in most societies are driven by consumerism and have been reluctant to explore the promise of simplicity because it threatens the engine of economic growth that is their lifeblood.

THREE VIEWS OF SIMPLICITY

I find it ironic that a lifeway of simplicity that can take us into an opportunity-filled future is often portrayed in the mass media as a primitive or regressive approach to life that turns its back on progress. Here are three major ways that I see the idea of simplicity presented in today's popular media:

1. *Crude or Regressive Simplicity:* The mainstream media often shows simplicity as a path of regress instead of progress. Simplicity is frequently presented as antitechnology and anti-innovation, a backward-looking way of life that seeks a romantic return to a bygone era. Regressive simplicity is often portrayed as a utopian, back-to-nature movement, with families leaving the stresses of urban life in favor of living in the woods, or on a farm, or in a recreational vehicle, or on a boat. This is a stereotypical view of a crudely simple lifestyle—a throwback to an earlier time and more primitive condition—with no indoor toilet, no

phone, no computer, no television, and no car. No thanks! Seen in this way, simplicity is a cartoon lifestyle that seems naïve, disconnected, and irrelevant— an approach to living that can be easily dismissed as impractical and unworkable. Regarding simplicity as regressive and primitive makes it easier to embrace a "business as usual" approach to living in the world.

2. *Cosmetic or Superficial Simplicity:* In recent years, a different view of simplicity has begun to appear—a cosmetic simplicity that attempts to cover over deep defects in our modern ways of living by creating the appearance of meaningful change. Shallow simplicity assumes that green technologies—such as fuel-efficient cars, fluorescent lightbulbs, and recycling—will fix our problems, give us breathing room, and allow us to continue to behave pretty much as we have in the past without requiring that we make fundamental changes in how we live and work. Cosmetic simplicity puts green lipstick on our unsustainable lives to give them the outward appearance of health and happiness. A superficial simplicity gives a false sense of security by implying that small measures will solve great difficulties and allow us to continue along our current path of growth for decades or more.

3. *Deep or Conscious Simplicity:* Occasionally presented in the mass media and poorly understood by the general public is a conscious simplicity that represents a deep, graceful, and sophisticated transformation in our ways of living—the work we do, the

transportation we use, the homes and neighborhoods in which we live, the food we eat, the clothes we wear, and much more. A sophisticated and graceful simplicity seeks to heal our relationship with the Earth, with one another, and with the sacred universe. Conscious simplicity is not simple. This is a lifeway that is growing and flowering with a garden of expressions. Deep simplicity fits aesthetically and sustainably into the real world of the twenty-first century.

Few people would voluntarily go through the difficulty of fundamentally restructuring their manner of living and working if they thought they could just tighten their belts and wait for things to return to "normal." A majority of people will shift their ways of living only when it is unmistakably clear that we must make dramatic and lasting changes. Are we there yet? Has the world reached a point of no return and crossed a threshold where a shift toward the simple prosperity of green lifestyles is the new "normal"?

WHAT KIND OF SIMPLICITY FITS OUR WORLD?

Although human societies have confronted major challenges throughout history, our era is unique. We will explore the tipping point our world has reached in much greater detail in Chapter 5. To summarize, here are four world-changing trends that illustrate what an exceptional time we are living in now—and how our lives will be profoundly different in the near future:

- **PEAK OIL** Only one time will we use up all of the world's reserves of oil. We have already used up roughly half of all the oil—the half that is easiest and cheapest to get—and global demand is skyrocketing. The price of oil will escalate and depress the global economy until the world can shift to renewable energy sources.

♦ **CLIMATE CHANGE** Only one time will we melt the world's ice caps and glaciers and radically destabilize the planet's climate. We are creating a new Earth for future generations and risking monumental crop failures and famines in this generation.

♦ **OVERPOPULATION** Only one time will we so unconsciously overpopulate the Earth, with our billions of people overwhelming the capacity of our land, water, and air ecosystems to regenerate themselves.

♦ **SPECIES EXTINCTION** Only one time will we cause the extinction of a third or more of all animal and plant species. The integrity of the web of life is one of the clearest measures of the health of the Earth. We are destroying large portions of the biosphere and putting at risk the very foundations of our existence.

Never before has the human family been on the verge of devastating the Earth's biosphere and crippling its ecological foundations for countless generations to come. The circle has closed and there is no escape. The Earth is a single, tightly interconnected system. Both the natural ecology of the Earth and the social ecology of human relations are being placed at profound risk. We confront far more than individual "problems." We are moving into an intertwined, world systems crisis involving every aspect of life. In this generation we meet fundamental questions head-on: Who are we? What kind of journey are we on as a human community? How are we to live together on this increasingly small Earth?

Our condition of planetary systems crisis should not be a surprise. As early as 1992 more than sixteen hundred of the world's senior scientists, including a majority of the living Nobel laureates in the sciences, signed an unprecedented "Warning to Humanity." In this historic statement, they declared that "human beings and the natural world are on a collision course . . . that may so alter the living world that it will be unable to sustain life in the manner that we know." This is their conclusion:

> We, *the undersigned senior members of the world's scientific community, hereby warn all humanity of what lies ahead.* A great change in our stewardship of the earth and the life on it is required, if vast human misery is to be avoided and our global home on this planet is not to be irretrievably mutilated.[2] *[emphasis added]*

What kind of "stewardship" fits our emerging world? When we consider the powerful forces transforming our world—climate change, peak oil, water and food shortages, species extinction, and more—we require far more than either crude or cosmetic changes in our manner of living. If we are to maintain the integrity of the Earth as a living system, we require deep and creative changes in our overall levels and patterns of living and consuming. *Simplicity is not an alternative lifestyle for a marginal few. It is a creative choice for the mainstream majority, particularly in developed nations.* If we are to pull together as a human community, it will be crucial for people in affluent nations to embrace a deep and sophisticated simplicity as a foundation for sustainability. *Simplicity is simultaneously a personal choice, a community choice, a national choice, and a species choice.*

What does a life of conscious simplicity look like? There is no cookbook we can turn to with easy recipes for the simple life. The world is moving into new territory and we are all inventing as we go. For more than thirty years I've explored contemporary expressions of the simple life and I've found such diversity that the most useful and accurate way of describing this approach to living may be with the metaphor of a garden.

A GARDEN OF SIMPLICITY

To portray the richness of simplicity, here are eight different flowerings that I see growing in the "garden of simplicity." Although there is overlap among them, each expression of simplicity seems sufficiently distinct to warrant a separate category. These are presented in no particular order, as all are important.

1. *Uncluttered Simplicity:* Simplicity means taking charge of lives that are too busy, too stressed, and too fragmented. Simplicity means cutting back on clutter, complications, and trivial distractions, both material and nonmaterial, and focusing on the essentials—whatever those may be for each of our unique lives. As Thoreau said, "Our life is frittered away by detail. . . . Simplify, simplify." Or, as Plato wrote, "In order to seek one's own direction, one must simplify the mechanics of ordinary, everyday life."

2. *Ecological Simplicity*: Simplicity means choosing ways of living that touch the Earth more lightly and that reduce our ecological impact on the web of life. This life-path remembers our deep roots with the soil, air, and water. It encourages us to connect with nature, the seasons, and the cosmos. An ecological simplicity feels a deep reverence for the community of life on Earth and accepts that the nonhuman realms of plants and animals have their dignity and rights as well.

3. *Family Simplicity*: Simplicity means placing the well-being of one's family ahead of materialism and the acquisition of things. This expression of green living puts an emphasis on providing children with healthy role models living balanced lives that are not distorted by consumerism. Family simplicity affirms that what matters most in life is often invisible—the quality and integrity of our relationships with one another. Family simplicity is also intergenerational—it looks ahead and seeks to live with restraint so as to leave a healthy Earth for future generations.

4. *Compassionate Simplicity*: Simplicity means feeling such a strong sense of kinship with others that, as Gandhi said, we "choose to live simply so that others may simply live." A compassionate simplicity means feeling a bond with the community of life and being drawn toward a path of cooperation and fairness that seeks a future of mutually assured development for all.

5. *Soulful Simplicity*: Simplicity means approaching life as a meditation and cultivating our experience of direct connection with all that exists. By living simply, we can more easily awaken to the living universe that surrounds and sustains us, moment by moment. Soulful simplicity is more concerned with consciously tasting life in its unadorned richness than with a particular standard or manner of material living. In cultivating a soulful connection with life, we tend to look beyond surface appearances and bring our interior aliveness into relationships of all kinds.

6. *Business Simplicity*: Simplicity means that a new kind of economy is growing in the world, with healthy and sustainable products and services of all kinds (home-building materials, energy systems, food production, transportation). As the need for a sustainable infrastructure in developing nations is being combined with the need to retrofit and redesign the homes, cities, workplaces, and transportation systems of developed nations, it is generating an enormous wave of green business innovation and employment.

7. *Civic Simplicity*: Simplicity means that living more lightly and sustainably on the Earth requires changes in every area of public life—from public transportation and education to the design of our cities and workplaces. The politics of simplicity is also a media politics, as the mass media are the primary vehicle for reinforcing—or transforming—the mass conscious-

ness of consumerism. To realize the magnitude of changes required in such a brief time will require new approaches to governing ourselves at every scale.

8. *Frugal Simplicity*: Simplicity means that, by cutting back on spending that is not truly serving our lives, and by practicing skillful management of our personal finances, we can achieve greater financial independence. Frugality and careful financial management bring increased financial freedom and the opportunity to more consciously choose our path through life. Living with less also decreases the impact of our consumption upon the Earth and frees resources for others.

As these eight approaches illustrate, the growing culture of simplicity contains a flourishing garden of expressions whose great diversity—and intertwined unity—are creating a resilient and hardy ecology of learning about how to live more sustainable and meaningful lives. As with other ecosystems, it is the diversity of expressions that fosters flexibility, adaptability, and resilience. Because there are so many pathways into the garden of simplicity, this self-organizing movement has enormous potential to grow.

MANY NAMES FOR A SHARED UNDERSTANDING

The movement toward simplicity is part of a "leaderless revolution" under way around the world. The revolution's principal concerns are building a sustainable future with the Earth, a harmonious relationship with one another, and a sacred relationship with nature and the universe. This is a self-organizing movement in which people are consciously taking charge of their lives. It is a promising demonstration of people taking responsibility for how their lives connect with the Earth and the future. Many of these pioneers have been working at the grassroots level for several decades, often feeling alone, not realizing that scattered through society, numbering in the millions, are others like themselves.

In his book *Blessed Unrest*, Paul Hawken describes the largely invisible rise of the world's largest movement working for environmental health and social justice, involving more than a million organizations around the world. He writes:

> *Across the planet, groups ranging from ad hoc neighborhood associations to well-funded international organizations are confronting issues like the destruction of the environment, the abuses of free-market fundamentalism, social injustice, and the loss of indigenous cultures. They share no orthodoxy and follow no single charismatic leader, yet they are organizing from the bottom up and coalescing into larger networks to achieve their goals—most urgently, ecological sustainability.*[3]

Because a concern for ecological sustainability is so widely shared, there are many ways of describing more sustainable and meaningful ways of living. There is no special virtue to the phrase "voluntary simplicity." Because this is a leaderless revolution in living, people are inventing as they go—including inventing words and phrases to characterize their approach to living.

Here are ten phrases in common use that offer an alternative to "voluntary simplicity":

- Green lifeways

- Earth-friendly living

- Soulful living

- Simple living

- Sustainable lifestyles

- Living lightly

- Compassionate lifeways

- Conscious simplicity

- Earth-conscious living

- Simple prosperity

To reflect this diversity of labels, I have made a point of using different phrases throughout the book that provide an alternative to "voluntary simplicity." Whatever we call this approach to living, a grassroots movement with many names is growing around the world, with three overriding and intertwined concerns—how are we to live sustainably on the Earth, in harmony with one another, and in communion with the universe?

MISCONCEPTIONS ABOUT THE SIMPLE LIFE

Four misconceptions about the simple life are so common that they deserve special attention. These are equating simplicity with poverty, with rural living, with ugly living, or with economic stagnation.

Simplicity Means Poverty Although some spiritual traditions have advocated a life of extreme renunciation, it is misleading and inaccurate to equate simplicity with poverty. My awakening to the harsh reality of poverty began on my father's farm in Idaho, where I worked with people who lived on the edge of subsistence. I remember one fall harvest when I was about ten years old in the early 1950s. We were harvesting a forty-acre field of lettuce, and a crew of twenty or so migrant laborers arrived to go to work. I still recall a family of three—a father, mother, and a daughter about my age—who drove their old Mercury sedan down the dusty road into our farm. They parked in the field and, with solemn faces, worked through the day doing piece labor—getting paid for the number of crates of lettuce they filled. At the end of the day they received their few dollars of wages as a family, earning roughly sixty-five cents an hour between them. That evening I returned to the fields with my father to check on the storage of the crates of lettuce and found the family parked at the edge of the field, sitting against the side of their car, and eating an evening meal that consisted of a loaf of white bread, a few slices of lunch meat, and a small jar of mayonnaise. I wondered how they managed to work all day on such a limited meal, but asked no questions. When I arrived for work the following morning, they got out of their car, where they had slept the night, and began working another day. After they had repeated this cycle for three days, the harvest was finished and they left. This was just one of innumerable personal encounters with poverty.

Over the next fifteen years, as I worked in the fields each summer, I gradually came to realize that most of the people working beside me did not know whether, in another week or month, their needs for food and shelter would continue to be met by their meager salary. As I worked side by side with these fine people, I saw that poverty has a very human face—one that is very different from "simplicity." Poverty is involuntary and debilitating, whereas simplicity is voluntary and enabling. Poverty is mean and degrading to the human spirit, whereas a life of conscious simplicity can have both a beauty and a functional integrity that elevates the human spirit. Involuntary poverty generates a sense of helplessness, passivity, and despair, whereas purposeful simplicity fosters a sense of personal empowerment, creative engagement, and opportunity. Historically those choosing a simpler life have sought the golden mean—a creative and aesthetic balance between poverty and excess. Instead of placing primary emphasis on material riches, they have sought to develop, with balance, the invisible wealth of experiential riches.

Simplicity Means Rural Living   In the popular imagination there is a tendency to equate the simple life with Thoreau's cabin in the woods by Walden Pond and to assume that people who opt for simplicity must live an isolated and rural existence. Interestingly, Thoreau was not a hermit during his stay at Walden Pond. His famous cabin was roughly a mile from the town of Concord, and every day or two he would walk into town. His cabin was so close to a nearby highway that he could smell the pipe smoke of passing travelers. Thoreau wrote that he had "more visitors while I lived in the woods than any other period of my life."[4] The romanticized image of rural living does not fit the modern reality, as a majority of persons choosing a life of conscious simplicity do not live in the backwoods or rural settings; they live in cities and suburbs. While green living brings with it a reverence for nature, that does not require moving to a rural setting. Instead of a "back to the land" movement, it is much more accurate to describe this as a "make the most of wherever you are" movement—and increasingly that means adapting ourselves creatively to a rapidly changing world in the context of big cities and suburbs.

Simplicity Means Ugly Living   The simple life is sometimes viewed as a primitive approach to living that advocates a barren plainness and denies the value of beauty and aesthetics. But while the Puritans, for example, were suspicious of the arts, many other advocates of simplicity have seen the arts as essential for revealing the natural beauty of things. Many who adopt a simpler life would surely agree with Pablo Picasso, who said that "art is the elimination of the unnecessary." Frank Lloyd Wright was an advocate of an "organic simplicity" that integrates function with beauty and eliminates the superfluous. In his architecture a building's interior and exterior blend into an organic whole, and the building, in turn, blends harmoniously with the natural environment.[5] Rather than involving a denial of beauty, simplicity liberates the aesthetic sense by freeing things from artificial encumbrances. From a spiritual perspective, simplicity removes the obscuring clutter and discloses the spirit that infuses all things.

Simplicity Means Economic Stagnation   Some worry that if a significant number of people simplify their lives it will reduce demand for consumer goods and, in turn, produce unemployment and economic stagnation. While it is true that the level and patterns of personal consumption would shift in a society that values green living, an economy that embraces sustainability can flourish. Although the consumer and material goods sectors would contract, the service and public sectors (education, health care, urban renewal) would expand dramatically. When we look around at the condition of the world, we see a huge number of unmet needs: caring for the elderly, restoring the environment, educating illiterate and unskilled youth, repairing decaying roads and infrastructure, providing health care, creating community markets and local enterprises, retrofitting the urban landscape for sustainability, and many more. Because there are enormous numbers of unmet needs, there are equally large numbers of purposeful and satisfying jobs waiting to get done. The difficulty is that in many industrialized nations there is such an overwhelming emphasis placed on individual consumption that it has resulted in the neglect of work that promotes public well-being. There will be no shortage of employment opportunities in an Earth-friendly economy. In moving toward simpler ways of living and a needs-oriented economy that does not artificially inflate consumer wants, an abundance of meaningful and satisfying jobs will become available, along with the ways to pay for them.

It is important to acknowledge these stereotypes, because they make a simpler life seem impractical and unapproachable and thereby reinforce the feeling that nothing can be done to respond to our critical world situation. To move from denial to action, it is vital to have an accurate understanding of a lifeway of conscious simplicity and its growing relevance for the human journey.

GROWTH OF GREEN LIVING

When I did the research for the first edition of *Voluntary Simplicity* in the late 1970s, the world was a different place. Thirty years ago, humanity was still blessed with cheap oil, a stable climate, and a population that was moderate in size relative to the world's resources. This produced an era of easy abundance, and in this setting, the simple life looked like a path of regress—a needless turn away from the "good life." However, three decades later—with the end of cheap oil, a destabilized climate, crop failures, and famines combined with a massive and growing global population that overwhelms available resources—the context for understanding simplicity is fundamentally different. The easy abundance of the past is being replaced by forced frugality. In this new setting, green lifeways of sophisticated simplicity take on new meaning. Instead of being dismissed as regress into the past, they are being welcomed as a path for creative progress into the future.

A growing number of people are deciding not to wait for leadership from others; instead, they are empowering themselves to invent alternative approaches to living that are more sustainable and satisfying. Recall the words of former astronaut Edgar Mitchell as he was contemplating the Earth on his return from the moon: "Personal responsibility for the greater good must become the mark of an informed and conscious people." And indeed, in the United States and around the world, there is clear evidence that a shift is under way as people take personal responsibility for contributing to the well-being of the world.

Although leaderless, this self-organizing movement for sustainability is growing rapidly around the world. In the United States and a dozen or so other "postmodern" societies (including those of Europe, Japan, and Australia), a movement toward green living has grown from a minuscule subculture in the 1960s to a respected part of the mainstream culture in the early 2000s. Glossy magazines now sell the simple life and green living on newsstands across the United States, and it has become a popular theme on major television talk shows. Based upon three decades of research, I estimate that as of 2009, roughly 20 percent of the U.S. adult population, or approximately forty million people, are consciously crafting Earth-friendly or green ways of living. These lifeway pioneers are providing the critical mass of invention at the grassroots level that could enable the larger society to swiftly develop alternative ways and approaches to living.

These changes are not confined to the United States and Europe. Around the world, people are awakening to the sanity of simplicity as a path to sustainability. Global surveys show that majorities everywhere support environmental protection and human development, two key themes accompanying lifeways of simplicity.[6] Global surveys also show there is virtually worldwide citizen awareness that our planet is indeed in poor health and that there is great public concern for its future well-being.[7] It makes little difference whether people live in poorer or wealthier nations—everyone expressed nearly equal concern for the health of the planet.

THE CHOICE FOR SIMPLICITY

The circle has closed. The Earth is a single system and we humans have reached beyond its regenerative capacity. It is of the highest urgency that we invent new ways of living that are sustainable. The starting gun of history has already gone off and the time for creative action has arrived. With lifestyles of conscious simplicity, we can seek our riches in caring families and friend-ships, reverence for nature, meaningful work, exuberant play, social contribution, collaboration across generations, local com-munity, and creative arts. With conscious simplicity, we can seek lives that are rich with experiences, satisfaction, and learning rather than packed with things. With these new ingredients in the lives of our civilizations, we can redefine progress, awaken a new social consciousness, and establish a realistic foundation for a sustainable and promising future.

two

PIONEERS OF GREEN LIVING

Making the simple complicated is commonplace;
making the complicated simple, awesomely simple,
that's creativity.

—*Charles Mingus*

What do simpler, more sustainable ways of living feel like? What is the lived experience of shifting priorities from material to nonmaterial sources of satisfaction? Who are these people who want to slow down, lighten their impact on the Earth, and grow the quality of their relationships with the rest of life?

For the most part, these are unpretentious pioneers. Quietly and without fanfare, millions of people have been crafting ways of living that touch the world more lightly and compassionately. Without major media attention to publicly mark this progress, simpler ways of living have emerged largely unnoticed in many countries:

- A city dweller plants an intensive garden on his roof and volunteers to work in a community garden in an abandoned lot.

- A busy executive begins meditating to reduce stress and shifts away from the old business game of acquiring power and money and toward serving customers, community, and ecology.

- A suburban family insulates their home, buys a fuel-efficient car, recycles, and shifts their diet away from meat and highly processed foods.

- A lawyer learns carpentry as an alternative profession, shops for clothes in secondhand stores, and buys used books.

Although these may seem like inconsequential changes in response to the immense challenges facing our world, they are of great significance. The character of a society is the cumulative result of the countless small actions taken day in and day out, by millions of persons. Small changes that may seem unimportant in isolation are of transformative significance when adopted by an entire society. In this chapter we will see, through the living example of several hundred persons from all walks of life, that there are workable and satisfying responses to the urgent challenges we now face.

The power of living examples to teach was brought home to me a number of years ago. I was attending a conference with a number of leading thinkers who were exploring the concept of a transforming society. Although the meetings were of great interest and many grand pronouncements were made concerning the need for social change, I no longer remember anything that was said. However, I do remember having lunch with Elise Boulding—a devout Quaker, feminist, sociologist, and compassionate advocate of the need for nonviolent, though fundamental, social change. At the end of the first morning's discussion we emerged from conference rooms to encounter an enormous buffet heaped with fruits, cheeses, salads, meats, breads, and more. Having worked up a considerable appetite, I filled my plate and sat down next to Elise. She had, without comment or display, selected for her lunch an apple, a piece of cheese, and a slice of bread. I was surprised that she had chosen such a modest lunch when such a bountiful offering was available. I asked Elise how she felt, and she reassured me that she was feeling fine. But I was still puzzled. I persisted and asked why she had taken such a small helping. In a few quiet sentences she explained that she did not want to eat what others in the world could not have as well.

In this seemingly small incident, I encountered a practical expression of the compassionate awareness that our individual well-being is inseparable from the well-being of other members of the human family. It is interesting and instructive that I don't remember any of the ideas from that conference, but I do recall what Elise had for lunch that day years ago—and why. I saw that it is the example of each person's life, much more than his or her words, that speaks with the greatest power. Even the smallest action done with a loving appreciation of life can touch other human beings in profound ways. I also saw that we each carry within ourselves the capacity for directly sensing appropriate action as we make our way through this complex and often confusing world.

THE VOLUNTARY SIMPLICITY SURVEY

To give a window into the motivations, values, and experiences of those choosing a path of sophisticated simplicity, in this chapter I present the results of a survey conducted in the late 1970s. Importantly, as I describe at the end of this chapter, the value of these firsthand insights has not been diminished by time. There is an enduring wisdom and relevance in the comments and reflections offered by respondents.

As background, the "Voluntary Simplicity Survey" was developed in 1977 while I was working as a social scientist and futurist at a major think tank, SRI International. The previous year I had coauthored a business report on voluntary simplicity with a colleague and friend, Arnold Mitchell.[8] This report was published as an article in the *Co-Evolution Quarterly,* a journal whose readership included many persons who have adopted more ecological ways of living. Seeing the opportunity, I included a questionnaire with this article that asked people to describe their experiences.

In the eighteen months after this article was published, more than four hundred questionnaires and over two hundred letters totaling more than a thousand pages were received. Nearly every letter offered a revealing journey into the lives of an individual or family. Many letters were inspiring and empowering. Some were light and humorous. A few were despairing. I read each letter at least four times. I was impressed with the compassion, humor, and courage of these individuals and families. Here is the background of these respondents:

+ People from all walks of life responded—lawyers, teachers, social workers, students, government bureaucrats, firemen, carpenters, factory workers, retired couples, white-collar workers, and others.

+ Responses were received from forty-two U.S. states as well as from several European countries, Canada, and Australia.

- A broad spectrum of ages was represented—from seventeen to sixty-seven. The average age was roughly thirty, and 75 percent were under the age of thirty-five.

- Nearly all respondents were white.

- Overall income levels tended to be somewhat lower than that of the general U.S. population.

- Most were highly educated—roughly 70 percent had completed college.

- A majority (56 percent) lived in cities and suburbs, 13 percent lived in smaller towns, and 32 percent lived in rural areas.

- Most grew up in relatively affluent homes (71 percent had a middle-class economic background and 22 percent had an upper-middle-class economic background).

- The average respondent had been choosing a simpler way of life for six years.

This background information revealed a number of important insights:

- People throughout the United States and other nations responded. The wide geographic distribution of respondents indicates this is not a regional phenomenon.

- The fact that a majority of persons were living in urban settings indicates this is not a predominantly "back to the land" movement.

- The fact that virtually all respondents were white and had childhood backgrounds of relative affluence indicates that the early adopters, or innovators, are not likely to come from those groups who grew up with poverty and discrimination.

- The fact that respondents had a moderately wide range of incomes suggests that, within limits, the manner in which one's income is used is as important as the size of income earned.

- The relatively high levels of education of members of this forerunner group indicates that they are not unskilled and poorly educated dropouts from society. Instead, they are well-educated and skilled persons who are searching for ways of living and working that are more sustainable and satisfying.

- The fact that the average length of time persons had been involved with this way of living was six years makes it clear that this is not a short-lived lifestyle fad.

I drew a representative sampling of comments concerning a broad range of topics from over one hundred of the more than two hundred letters received. In order to ensure the privacy of these individuals and still reveal some of the flavor of their lives, each quotation was identified solely by the person's sex, age, marital status, degree of urbanization of residence, and geographic locale.

WHAT IS VOLUNTARY SIMPLICITY?

To begin, how does this community—which practices what it preaches—define this way of life for themselves? Here is a sampling of responses that seem so fresh they could have been written yesterday:

> Voluntary simplicity has more to do with the state of mind than a person's physical surroundings and possessions.
>
> —*woman, twenty-three, single, small town, East*

> Simplicity meant I fit easier into the more ecological patterns, hence was more flexible, more adaptable, and ultimately more aware of the natural spiritual path before me; less nonsense in the way.
>
> —*man, thirty, single, small town, West*

> As my spiritual growth expanded and developed, voluntary simplicity was a natural outgrowth. I came to realize the cost of material accumulation was too high and offered fewer and fewer real rewards, psychological and spiritual.
>
> —*man, twenty-six, single, small town, South*

> I don't think of it as voluntary simplicity. I am simply going through a process of self-knowledge and self-realization, attempting to better the world for myself, my children, my grandchildren, and so on.
>
> —*woman, thirty-eight, married, suburb, East*

To me, voluntary simplicity means integration and awareness in my life.

—*woman, twenty-seven, single, city, South*

In realizing true identity (a continuing process), voluntary simplicity naturally follows.

—*man, age unspecified, living together, small town, West*

Ecological consciousness is a corollary of human consciousness. If you do not respect the human rights of other people, you cannot respect the Earth. The desires for material simplicity and a human-scale environment are results of an ecological consciousness.

—*man, twenty-six, single, small town, West*

I can't call us living simply but rather living creatively and openly.

—*man, thirty-one, married, rural, West*

Voluntary simplicity is not poverty, but searching for a new definition of quality—and buying only what is productively used.

—*man, twenty-eight, married, rural, West*

We laugh that we are considered a "poverty" family, as we consider our lives to be rich and full and completely rewarding—we are living in harmony with everything. I know for myself the source of "richness" or "poverty" comes from within me.

—*woman, forty-one, married, rural, West*

While each person had his or her own, unique description of the simple life, a common theme was a concern for the psychological and spiritual aspects of living. Still, the diversity of definitions was important, as many said they were wary of rigid views that would be dogmatically applied and result in a self-righteous, "simpler-than-thou" attitude.

Another concern expressed in the letters was that living more simply would not be seen as a fundamental shift in one's way of life and instead would be trivialized by the mass media and portrayed as only a cosmetic and superficial change. Here are two illustrative comments:

> This is a country of media hype, and [simple living] is good copy. The media is likely to pick up on it . . . and create a movement. I hope they won't. The changes we're talking about are fundamental and take lots of time. . . . If it is made into a movement, it could burn itself out. I hope it spreads slowly. This way the changes will be more pervasive. Voluntary simplicity is the kind of thing that people need to discover for themselves.
>
> —*man, twenty-seven, single, rural, Midwest*

> Make a movement out of a spontaneous tendency like voluntary simplicity and the best aspects of it (and the individuals) will elude you.
>
> —*woman, thirty, living together, rural, West*

Many of the persons experimenting with simpler ways of living said they did not view themselves as part of a conscious social movement. Instead they were acting on their own initiative to bring their lives into greater harmony with the needs and realities of the world. This reinforces a conclusion of the preceding chapter that this is a leaderless movement in which people are taking personal responsibility for the greater good of the Earth and the rest of life.

WHY CHOOSE SIMPLICITY?

Why would an individual or couple adopt a way of life that is more materially frugal, ecologically oriented, inner-directed, and in other ways removed from the materialism of much of Western society? Here is a sampling of the reasons given:

> I believe in the imminent need for the skills and resources I am developing now. I am not sure how it will come about, whether economic collapse, fuel exhaustion, or natural disaster, but whichever it is, I (and my family) will need all of whatever self-sufficiency I or we can develop.
>
> —*man, twenty-nine, married, rural, West*

> I believe voluntary simplicity is more compassionate and conducive to personal and spiritual growth. I live this way because I am appalled that half the planet lives in dire poverty while we overconsume. And people think they are "Christian." I think it is "spiritual" to make sure that everyone has adequate food, shelter, and clothing and to take care of the planet.
>
> —*woman, twenty-five, married, big city, Midwest*

> I sincerely believe that voluntary simplicity is essential to the solution of global problems of environmental pollu-

tion, resource scarcity, socioeconomic inequities and existential/spiritual problems of alienation, anxiety, and lack of meaningful lifestyles.

—*man, thirty-two, married, suburb, South*

I have less and less to blame on other people. I am more self-reliant. I can both revel in the independence and be frustrated by my shortcomings—but I get to learn from my own mistakes. Each step is progress in independence; freedom is the goal.

—*man, twenty-six, married, small town, East*

The main motivation for me is inner spiritual growth and to give my children an idea of the truly valuable and higher things in this world.

—*woman, thirty-eight, single, small town, East*

I feel more voluntary about my pleasures and pains than the average American who has his needs dictated by Madison Avenue (my projection of course). I feel sustained, excited, and constantly growing in my spiritual and intellectual pursuits.

—*woman, thirty, living together, rural, West*

Why simplicity? I see it as the only moral, economic, rational, humanistic goal. Besides, it's fun.

—*man, twenty-three, single, small city, Midwest*

It was the injustice and not the lack of luxury during the Great Depression that disturbed me. I took up this way of life when I was seventeen. I remember choosing this simplicity—not poverty—because: it seemed 1) more just in the face of deprivation—better distribution of goods; 2) more honest—why take or have more than one needs; and 3) much freer—why burden oneself with getting and caring for just "things" when time and energy could be spent in so many other more interesting and higher pursuits? 4) But I wanted a simplicity that would include beauty and creativity—art, music, literature, an aesthetic environment—but simply.

—woman, sixty, married, suburb, West

Our interest in simpler living dates to overseas tours with the U.S. embassy in underdeveloped nations—we know firsthand what the problems are.

—woman, sixty-one, married, suburb, East

I felt the values involved in consumerism to be false, useless, and destructive. I prefer to appear as I am. People are complex enough to understand without excess trappings. I was also influenced by the values of the feminist and ecological movements.

—woman, twenty-five, married, suburb, South

Increasing my self-sufficiency seemed the only honest way to effectively make my feelings, actions, and life congruent.

—man, twenty-seven, married, rural, East

It is a highly rewarding way to live. It forces you into a relationship with a basic reality. . . . It also forces you to deal with some direct anxieties and rely on and be thankful to a benevolent deity. It succinctly points out your frailty and clearly delineates your dependencies. It also reinforces your strength and independence.

—married couple, thirty-seven and thirty-two, rural, West

I wanted to remove my children from the superficial, competitive (East Coast) value system. Wanted a family venture to draw us closer and a community that was stable. Also wanted to provide the children with a learning experience that exposes them to alternatives to the "rat race" system, plus I wanted out from the typical pressures of maintaining material acquisitions that were meaningless to me.

—woman, thirty-six, single, rural, West

Here are the most common reasons given for choosing to live more lightly:

- To find a healthy balance between one's inner experience and its outer expression in work, consumption, relationships, and community.

- To search for a workable and meaningful alternative to the emptiness of a society obsessed with material consumption and display.

- To provide one's children with more humane value systems and life experiences that are appropriate to the emerging world they will have to live in.

- To find a much higher degree of independence and self-determination in a mass society of alienating scale and complexity.

- To establish more cooperative and caring relationships.

- To acknowledge and, in small but personally meaningful ways, begin to reduce the vast inequities between the rich and the poor around the world.

- To cope in a personal manner with environmental pollution and resource scarcity.

- To foster nonsexist ways of relating.

- To develop the personal skills and know-how to survive a time of severe economic and social disruption.

- To create the personal circumstances of life in which one's feelings, thoughts, and actions can come into alignment.

THE PATH TO SIMPLE LIVING

What is the pathway from consumer to conserver? Is green living or simple living chosen abruptly or does it evolve gradually? Here is a sample of comments:

We are moving toward a life of greater simplicity from within, and the external changes are following—perhaps more slowly. We are seeking quality of life—and a path with heart.

—woman, age unspecified, married, suburb, West

Voluntary simplicity must evolve over a lifetime according to the needs of an individual. . . . The person must grow and be open to new ideas—not jump on a bandwagon, but thoughtfully consider ideas and see how they relate to oneself.

—woman, twenty-one, single, small city, Midwest

[Voluntary simplicity is] a thorough lifestyle that only works through full commitment and takes many years to grow into.

—man, twenty-nine, married, small town, West

Simplicity began unfolding in my life as a process. It was an inarticulate but seemingly sensible response to emerging situations—and one response after another began to form a pattern, which you identify as voluntary simplicity. For me simplicity was the result of a growing awareness plus a sense of social responsibility.

—man, thirty-one, married, small town, West

To me, voluntary simplicity as a lifestyle is not something you take up in one moment, but occurs over a period of time due to: (1) consciousness raising; (2) peer group support; (3) background; (4) inner-growth interest; and many other factors. My wholehearted commitment to a certain spiritual path finds outer expression in a simple, gentle, humane lifestyle.

—woman, twenty-eight, single, suburb, West

It wasn't a slam-bang, bolt-from-the-blue, overnight change. I'm still growing and learning. The most important goal I have is inner development with a good blend of living with the here and now on this planet.

—*man, twenty-five, married, small town, East*

I consciously started to live simply when I started to become conscious. . . . Living simply is in the flow of things for me.

—*man, thirty, married, rural, East*

My ideas and my practice of voluntary simplicity have been and I hope will continue to be a gradual process of evolution and growth. From early adolescence on I tended to prefer simplicity.

—*woman, twenty-one, single, small city, Midwest*

Various flirtations with yoga, meditation, drugs, and radical politics gave me exposure to, and some personal experience with, "inner growth" possibilities. I began living my life freely, following no preconceived roles, and gradually discovered my overriding interest in quality: the environment, life, the universe.

—*man, twenty-nine, married, small town, East*

I became interested in simplicity . . . primarily because of ecological concerns. However, since then my interest has become concentrated more on metaphysics, self-realization, and so on, with the same end results.

—*woman, thirty-eight, married, big city, Midwest*

Voluntary simplicity is an individual thing. . . . It has to be something that springs from the heart because it was always there, not something you can be talked into by persuasive people, or something that is brought on by financial necessity. . . . This is not something we do because we want to be different, or because we're rebellious to convention, but because our souls find a need for it.

— *woman, thirty, married, big city, Canada*

Overall the journey into this way of life seems to be a relatively slow, evolutionary process, one that unfolds gradually over a period of months and years. The initial stages are a time of exploring and moving back and forth between traditional and innovative patterns of living and consuming. Gradually a person or family may find they have made a number of small changes and acquired a number of slightly different patterns of perception and behavior, the sum total of which adds up to a significant departure from the contemporary way of life. One conclusion that I draw from these letters is that if change is too abrupt, it may not have the staying power to last. It seems better to move slowly and maintain a depth of commitment that can be sustained over the long haul.

The letters also reveal that the transition into this way of life is a personally challenging process. It may be accompanied by inner turmoil and feelings of uncertainty, self-doubt, anxiety, despair, and more. Life changes of this dimension are seldom made without deep soul-searching. The support of friends, family, and work associates is of great importance for transforming a stressful and difficult process into a constructive and shared adventure.

The path toward simpler living is further revealed in people's descriptions of how they consciously altered their everyday lives:

I quit my forty-hour-a-week slavery and got a twenty-hour-a-week job that I love (working in a library). I started learning how to grow food in the city and make compost. I became conscious of what I was eating and how I was spending my money. I started learning to sew, mend, and shop secondhand, and I've stopped eating meat.

—*woman, twenty-three, married, small city, West*

I do not own anything more than I need. The things that I do own are selected on the basis of their utility, rather than their style or the fact that they are currently faddish. I attempt to make things last. . . . I am nursing my car past 100,000 miles. I am doing political work, notably in opposition to nuclear power. . . . I am planning to build my own house, and the plans include small-scale technology aimed at promoting self-sufficiency, such as passive solar design, a greenhouse, composting toilets, windmills, and so on.

—*man, thirty, single, suburb, West*

Changes include smaller house, wear clothes longer (except when in court; I'm a trial lawyer, feel it necessary to "play the role" when actually engaged in formal professional activity), recycle and buy secondhand when possible, bike and hike . . . live with a nice lady, have more time for children. . . . Human relationships, though fewer, are closer.

—*man, forty-two, living together, rural, East*

. . . quit smoking, stopped eating meat, now run about eight miles a week, stopped shaving legs, stopped using scented products, stopped buying stylish haircuts, buy less

clothing, buy looser, freer clothing, regularly take vita-
mins, gave away a lot of things, eat 90 percent more fruits
and vegetables, meditate, walk a lot, read humanistic psychology, study Sufism, feel strong affinity for all animals,
weave, write.

—woman, thirty-three, single, big city, West

I am doing what Bucky Fuller calls doing more with less.
He also speaks of education as the process of "eliminating
the irrelevant," dismissing all that is not furthering our
chosen articulation of value—eliminating wasteful speech
as well as costume, dietary habits as well as information
addictions that do not further the evolution into that simple (not to say "noncomplex" but only "noncomplicated")
life of adaptive progress to more and more diversified environments.

—man, twenty-seven, single, small city, East

I recycle cans, bottles, and newspapers. We're very careful
with water. . . . I buy used and handmade things as much
as possible. . . . We've always been frugal in the way we
furnish our house. We've never bought on time, which
means we buy fewer things. We wear other people's hand-
me-downs and we buy used furniture when possible. . . . A
large percentage of our spending goes for classes (music,
dance, postgraduate courses for my credential), therapy,
and human-potential experiences.

—woman, forty-seven, married, big city, West

We have a car but seldom use it, preferring to use bicycles
because of a car's pollution and energy consumption. I am

not into fashion and attempt to wear things till they are worn out—buy mostly serviceable work-type clothes, sometimes secondhand from friends. . . . Am vegetarian . . . belong to a food cooperative . . . everybody contributes four hours per month to working at the store/restaurant. This co-op forms an important hub in our community for most alternate social and spiritual activities. Learning about gardening . . . buy tools and appliances that are durable . . . avoid buying plastic and aluminum whenever possible. No throwaways . . . I attempt to use my buying power politically . . . strongly support appropriate technology . . . strongly motivated to understand myself and others—involved in meditation for awareness. . . . The spiritual-search component is the major driving force in my life. . . . I always try to acquire new self-help skills: sewing, car repair, and so forth.

—*woman, twenty-five, living together, city, Canada*

Greater simplicity frees time, energy, and attention for personal growth, family relationships, participation in compassionate causes, and other meaningful activities.

INNER GROWTH AND SIMPLICITY

The term "inner growth" refers generally to a process of learning a natural quietness of mind and openness of heart that allows our interior experiencing to become apparent to us. It is a process of going behind our day-to-day labels and ideas about who we think we are in order to make friends with ourselves and the world. Here are some illustrative comments regarding the importance of conscious simplicity to inner growth:

I don't believe a person can make the commitment necessary to maintain this lifestyle without a spiritual-
psychological motivation.

—man, thirty-three, single, small town, South

It seems to me that inner growth is the whole moving force
behind voluntary simplicity.

—woman, thirty-eight, single, suburb, West

My life is suffused with joy, and that transforms even the
ordinary day-to-day unpleasantries that come along. I've
had a lot of years of growth pains and for a long time I got
lost in the pain and suffered; but I've learned to let go more
easily now, and even the hard stuff that comes along
doesn't overpower the joy.

—woman, thirty-five, married, rural, West

My husband and I . . . feel that our inner growth at this
time is a daily way of life—tailored to our own particular
situations. We know from life that one does not step in the
same stream twice . . . thus, one keeps observing and relating to situations as they flow.

—woman, fifty-five, married, small town, West

I consider the whole picture one of positive personal
growth that I inadvertently (luckily?) set in motion through
trial and error and suffering. And now I sense a momentum established that I could not now "will" to halt.

—woman, thirty-three, single, big city, West

I consider every moment a chance for growth. To pay attention and learn is my meditation.

—*man, twenty-seven, married, rural, East*

In the opinion survey, people were asked if they were "now practicing or actively involved with a particular inner-growth process." The results, presented below, are striking. (Keep in mind that the percentages can exceed 100, since a person was asked to check those that apply and it was not uncommon for a person to be engaged in more than one growth process.)

- Meditation (e.g., Zen, Transcendental Meditation) 55%

- Other (e.g., biofeedback, intensive journaling) 46%

- Human potential (e.g., encounter groups) 26%

- Traditional religion (e.g., Catholicism, Judaism) 20%

- None 12%

- Psychoanalysis (e.g., Freudian, Jungian) 10%

A number of key findings emerge from this question. First, an overwhelming majority of the respondents indicated they were presently involved with one or more inner-growth processes. Second, more than half of the respondents indicated involvement with some form of meditative discipline—in fact, meditation appears to be the most prominent inner-growth process correlated with a life of conscious simplicity. Third, a near majority were involved with other, often highly personalized, inner-growth processes. The category "Other," as described in the letters, runs the gamut from keeping an intensive journal to yoga, jogging, and many things in between. Fourth, although traditional religion is important, it is not the primary focus for a group that is so strongly psychologically and spiritually oriented.

Inner growth is even more important to a path of voluntary simplicity than the foregoing statistics suggest. A number of persons who declined to identify themselves as being involved in any particular growth process explained why with comments such as these:

> I don't know what label to put on my inner-growth process. . . . I get high on a beautiful sunrise, a night when I sit alone on a chunk of granite and gaze up at billions of stars. . . . So what would you call it? I like to sit alone on a rock and just open my mind up to everything—would you call it meditation? And I do believe in God and Christ, but for sure am not into traditional religion.
>
> —*man, twenty-four, single, rural, Midwest*

> I am not actively involved in any particular growth process, but I'm always aware of the Spirit and how I'm part of it— gleaning what I can from traditional religions, yoga, and the insights provided to a clearer head since I've begun living more simply.
>
> —*man, twenty-two, single, rural, South*

> I'm 75 percent spiritually oriented, though I haven't found a particular "path" outside of living simply and with love and searching.
>
> —*woman, eighteen, single, suburb, East*

I practice meditation and have a strong spiritual emphasis in my life that draws mainly from Christianity and Buddhism but also all other religions, and I do not practice any of them in any strict sense. I would call myself very strongly religious, but independent.

—woman, twenty-five, married, big city, Midwest

. . . tried Zen, but it was wrong pace—but work constantly on self in a small, personalized manner.

—man, twenty-four, single, rural, West

I do not deny taking ideas from many sources, but the result is fairly eclectic. The fact that most people I know do not fit me into a category seems to support this. So, my process would be individualistically developed inner growth: that which suits me personally.

—woman, twenty-six, single, big city, West

The letters indicate a basic dilemma: When one's entire life is the inner-growth process, how can any one piece be singled out? The category "None" thus includes an undetermined, but substantial, number of persons for whom the inner-growth process is really "all."

Some objected, saying they felt the category "Inner Growth" was too limiting and that activities from political action and artistic creation to running should be considered legitimate vehicles for inner growth:

I am a radical and a feminist and believe that leading a simpler lifestyle and continuing political struggle is a form of personal growth left out of your analysis.

—man, twenty-nine, single, rural, West

The inner growth you write of may be artistic, scientific, or social action, not just "spiritual."

—*man, twenty-eight, single, small city, West*

Athletics is played down unjustifiably. For me, human potential has a physical as well as a spiritual meaning. Distance running produces a lot of the same vibrations that meditation evokes.

—*man, twenty-nine, married, small city, West*

In short, just as there is no single "right" way to outwardly live more simply, so, too, there is no single "right" way to engage in the process of interior growth.

Overall, this survey revealed an important insight: *The single most common factor among respondents was an emphasis on inner growth and being awake to the miracle of life.* Living more consciously seems to be at the core of a path of simplicity and, in turn, makes it clear why this way of life is compatible with Christianity, Buddhism, Hinduism, Taoism, Sufism, Zen, and many more traditions. Simplicity fosters a more conscious and direct encounter with the world. From a more intimate encounter with life there naturally arise the powerful experiences at the heart of all the world's great spiritual traditions.

THE IMPORTANCE OF RELATIONSHIPS

The involvement with inner growth may suggest an inward turning away from worldly relationships. To the contrary, it is the very deepening of insight through the inner quest that reveals the entire world as an intimately interconnected system. In an ecological reality—where everything is related to, and connected with, everything else—the quality and integrity of interpersonal relationships are of great concern. Or, as Martin Buber said, "All real living is meeting." This, too, is reflected in the letters:

> We value our relationships more highly than anything else.
> —*man, twenty-nine, married, small city, West*

> I feel good about everyday things, place more value and get much satisfaction from interpersonal relationships.
> —*woman, nineteen, single, suburb, West*

> I feel this way of life has made my marriage stronger, as it puts more accent on personal relationship and "inner growth."
> —*man, twenty-seven, married, rural, East*

> We are intensely family-oriented—we measure happiness by the degree of growth, not by the amount of dollars earned.
> —*man, thirty-one, married, small town, East*

> Satisfactions—growth through relationships with others and continuing personal contacts—the circle expands as more interests develop.
> —*man, twenty-nine, living together, small town, East*

This is not to portray a utopian situation. These are very human people. Many have confronted the fact that movement toward a fundamentally different way of life—involving change in both inner and outer worlds—can be challenging for relationships, even when there is a joint commitment to change. The varied nature of the stresses that may accompany movement toward voluntary simplicity is revealed in the letters:

> We see some friends just grow old and tired of fighting [to maintain this way of life], especially when not backed up by solid love connections.
>
> —*man, thirty-four, married, rural, Canada*

> I expect to [live more simply] in the future as I become more ingenious about dealing with differences in lifestyle preferences of my husband and myself and more firm about saying no to outside demands.
>
> —*woman, fifty-seven, married, suburb, West*

> It is sometimes difficult or frustrating to fully live a life of simplicity because my partner is not as fanatically committed as I am to personal-growth exploration, non-consumerism, and conservation. Satisfactions, however, are innumerable—I feel better physically, spiritually, and mentally.
>
> —*woman, twenty-eight, married, suburb, West*

> I feel that I am alone with no support. Oh, some of my friends think that the things we are doing are admirable in a patronizing sort of way. Most people just think we're a little "nuts." I need some support. My wife is a somewhat reluctant supporter.
>
> —*man, thirty-six, married, big city, West*

My father and mother are supportive of our lifestyle; his parents are critical and don't understand—they are very money-oriented. The kids sometimes worry that we are poor and compare themselves with their more conventional classmates.

—*woman, twenty-nine, married, rural, West*

My choice of lifestyle, I feel, cannot infringe so much on my almost grown children's desires that it makes them miserable and rebellious. If they had all been toddlers when we began living this way, it would have been easier.

—*woman, thirty-six, single, rural, East*

Clearly, movement toward simpler living has its tensions and stresses. For couples, the most frequently mentioned source of tension within a relationship arose when one partner was content with traditional patterns of living, consuming, relating, and working while the other partner wanted to explore nontraditional alternatives to one or more of them. For couples with children, the most frequently mentioned source of stress involved children feeling themselves in a double bind between a traditional way of life shared by their peer group and an emergent way of life being explored by their parents.

Despite these stresses, relationships were described overall as an important source of nurturance, love, encouragement, reality testing, and support. Often this support extended beyond one's immediate family or friends to include a larger community of persons with whom there was a sense of kinship and spiritual bonding.

SIMPLICITY AND COMMUNITY

Many persons in this poll reported they felt tolerated by the larger community and actively supported by a smaller group of friends and associates. This is not surprising when we consider that many began adopting this way of life at a time that most nations were not strongly supportive of a trend toward simpler living. Here is a sampling of the spectrum of experience reported in the letters:

As for community support—I get both jeered at and cheered at—and some people join in [with support].

—*woman, fifty-five, married, small town, West*

My motivations are understood by very few; I accept the responsibility for my actions.

—*man, twenty-seven, married, rural, East*

I am greatly enriched by my lifestyle but sometimes feel alienated from society because of it. I consider it takes strong determination to stick to [my] convictions; sometimes frustrating because there are so many American dreamers.

—*woman, twenty-seven, single, small town, West*

I thought that the rural community would be a mecca of higher consciousness and have generally a spiritual atmosphere. Wrong! I have a problem finding people I have enough in common with that I can communicate with, but when I do, it's really dynamite!

—*man, thirty-three, single, small town, South*

It is scary to live with less because for so long our society has said that money, possessions, and a career lead to security and happiness. I have a lot of support to make changes because of a tremendous community of people who are doing the same thing. The advantage is feeling more inner peace, increased self-acceptance, and community support.

—*woman, twenty-nine, single, big city, Midwest*

Probably we will lead lives of greater simplicity to the degree that others in greater numbers in the society also start to do so. Community support is important.

—*woman, thirty, married, big city, West*

In the 1970s, many of the pioneers of the simple life did not feel substantial support from the larger community or culture, but three decades later, mainstream values are shifting in favor of green living and conscious simplicity. Community support—so important in empowering individuals to move from sympathetic leanings to active involvement—is growing rapidly. As we move from an era of easy abundance to one of forced frugality, a life of sophisticated simplicity is an appealing alternative.

THE POLITICS OF COMPASSION

The political orientation of this group represents a break with mainstream politics. This is a strongly independent community of persons whose perspectives do not fit within the traditional spectrum of liberal and conservative politics.

Among the U.S. respondents, only 1 percent labeled themselves Republicans. Some 28 percent said they were registered Democrats, but many included comments such as "out of lack of choice" or "would switch to something better." Overall there was not a strong affiliation with either political party. Roughly 60 percent of the sample placed themselves either in the category of "Independent" or "Other," and 11 percent declined to give any political orientation. The truly experimental character of the politics of this group is reflected in their responses to the category "Other." This brought write-in responses such as: "cooperativist," "feminist," "nonviolent activist," "whatever works at the time," "Libertarian," "eclectic," "decentralist," "conservative anarchist," and "apolitical but the weather may change." What seems to bind this group together is not political ideology but an appreciation for the dignity and preciousness of all life. Virtually nowhere in the letters was there any reference to traditional political ideologies; instead there was a sense that a whole new perspective is emerging that bridges between left and right, East and West, inner and outer, rich and poor, masculine and feminine.

> Our system causes us to be very political. . . . We are not a centralized movement. . . . Our concerns are for each other's survival. We communicate better than government and business. The "movement" is in reality a conscious choice of individuals—not a centralized program. . . . We are passing free ideas around, becoming independent through cooperation—young, old, brilliant, salt of the earth, revolutionary, conservative. We are the shakeout of the system, about to show the system how to improve itself. We communicate by learning and teaching. Call it the People's University, not a movement.
>
> *—man, thirty-five, married, suburban, East*

> I am spending more and more of my time in various forms of community building and feel there are no political organizations, ideologies, or labels with which I am continually comfortable—except maybe voluntary simplicity.
>
> *—man, thirty-eight, single, small city, West*

Different aspects of [the simple life] have the potential of touching small parts of everyone's lifestyle, and it is this feature that I feel gives hope to the movement. Voluntary simplicity has the potential of working on people like a chain reaction once the first spark is lit. On the other hand, I also see great potential for ecological disaster, social and political unrest, and ruin. I make no bets; I live from day to day and do what I can.

—woman, twenty-eight, married, suburb, West

The fire ignited by combining lifestyle and politics is truly liberating and a joy to experience. Revolutions are made of such stuff.

—man, twenty-six, single, small town, West

I don't think our lives will be very satisfying if we are divided and live in fear of others, regardless of what we may be able to accomplish with our inner lives. Thus if we are seeking genuine simplicity, we will have to operate in a way that will help others to gain awareness of their higher self-interest in what is happening to our planet and to ourselves as human beings.

—woman, fifty-five, married, small town, West

For several reasons those choosing a simpler life seem more engaged politically than the general population. First, these individuals are taking control of their immediate life circumstances, and although this action is limited to their own lives, it is real and empowering. In small but cumulatively significant ways they are experiencing their own competence to make constructive changes, and this spills over into their lives as citizens.

Second, there is the realization that, even if they wanted to drop out, there is literally no place to escape the impact of a disintegrating global ecology. In an interdependent world there is not the option of being apolitical. Once we acknowledge the plight of the world, there is no place to go to retreat into our former ignorance.

Third, living with less is politically radicalizing. The survey revealed that those with higher incomes were more traditional in their political perspectives, whereas those with lower incomes were more innovative. When people choose to live closer to the level of material sufficiency, they are brought closer to the reality of material existence for a majority of persons on the planet. There is not the day-to-day insulation from material poverty that accompanies affluent living.

In various ways, many respondents said they did not want to drop out of politics, but rather to change their manner of participation. In general, people seemed more interested in local and global concerns than with national issues. Examples of local concerns were changing zoning codes to allow the use of innovative building designs and materials. Examples of global concerns were actions to stop the destruction of rain forests and to preserve endangered species. This is a political perspective that is primarily concerned with building a sustainable future for all life on the planet.

Where the traditional political perspective of Western industrial nations has tended to emphasize national interests, material concerns, and the use of violence in the conduct of foreign affairs, the politics of simplicity tends to emphasize planetary interests, a balanced regard for both material and spiritual development, and the importance of nonviolent means of resolving disputes. Other common themes in the letters were an emphasis on cooperation and sharing, self-reliance and neighborliness, human rights and social justice.

CONTRIBUTIONS OF THE FEMININE PERSPECTIVE

The women's movement has made important contributions to both the growth and the character of simple living in the modern era. However, an affinity for a feminist perspective is no guarantee of sympathy for a simpler life. Many women have broken free of traditional sexual stereotypes but continue to embrace other stereotypes of Western industrial societies—particularly those concerning material success and social status. Nonetheless a substantial majority of those choosing a life of deliberate simplicity expressed a strong affinity for the spirit of the feminist movement. Here are some illustrative comments from the letters:

> I am becoming tuned to these ideas and lifestyle changes, partly as a feminist who sees a need for more bridges from the new age to feminism. . . . Behind this, of course, the relationship of our alienation/destruction of our mother, the earth, is parallel to our alienation/control of our mothers, the women.
>
> —*woman, thirty, married, big city, Canada*

> I am a feminist—the women's movement gave me the strength and knowledge and confidence in myself and abilities to attempt this lifetime and lifestyle adventure. . . . I am especially supportive of women who want to "try their wings" and serve as an example of what a woman can do when she rejects prescribed roles and does "her own thing."
>
> —*woman, thirty-nine, living together, rural, East*

I took up voluntary simplicity after leaving my last (hopefully) male-dominated relationship, where I was supported financially by a man. A year of feminism and consciousness raising . . . convinced me that I would never be free to even know a man truly unless I could be free of dependency. . . . So supporting myself, seeing how I really wanted to spend my waking hours, coupled with the concept of right livelihood, ecological awareness, yoga study, all led in one direction.

—woman, thirty-four, single, rural, West

Another part of [simpler living] is more equally shared roles of men and women. Even though we have a child (which is often the determining factor in the traditional mother-stay-home, father-go-to-work family) we are attempting to not put ourselves in those roles. . . . Between us we share child care, meal fixing, cleaning, shopping, and money earning, while making sure our other needs are met (crafts, classes). For people without children this is even more the case.

—woman, twenty-two, living together, small town, West

It is a way of life that made sense to me—nonmaterialistic, focused on people, cooperative, nonsexist.

—woman, twenty-seven, married, big city, Midwest

Major changes include elimination of the "man is the chief breadwinner and boss" hang-up.

—man, thirty-eight, married, suburb, West

The feminist movement has contributed to the growth of simpler living in several ways. First, feminism, by its example, has encouraged people of both sexes to explore alternative ways of living and working. When persons or groups empower themselves to act in ways that move beyond traditional roles and expectations, it provides an example of cultural liberation that all can emulate and translate into their unique circumstances. The liberation of women from sexual stereotypes has relevance far beyond women and sexual roles—it is a significant example of cultural liberation that applies to many other limiting stereotypes of traditional Western industrial societies.

Second, in liberating men from the need to perpetuate their half of the polarity of sex-based roles, feminism offers both men and women the freedom to be more authentically themselves. This has important implications for the male-dominated orientation of Western industrial societies, where proof of "manhood" has often been equated with the ability to succeed in the material world. For many men consumption has served purposes far beyond that of meeting genuine material needs. High earnings and high consumption have been used as evidence of masculine competence, potency, and social status. With changes in male-female roles, other criteria of "success" can begin to emerge— criteria that are more balanced across both masculine and feminine qualities.

With greater balance between masculine and feminine qualities, our cultures would tend to become less aggressive, contain less disguised competition, be more receptive and open, have more supportive friendships, have a greater mixing of roles among men and women in accordance with innate interests and capacities, foster nurturance and caring for others to a greater degree, place greater value on feeling and emotion, express greater concern for unborn generations, and have a stronger sense of the intimate interrelatedness of life. This integration and balance is crucial. If the masculine orientation—with its competitive, aggressive, dissecting, and materialistic approach to living—continues to dominate our perceptions and actions as a culture, we will scarcely be able to live in relative peace with the rest of the life on this planet. If we are to become whole persons in a cohesive culture, we must consciously integrate more feminine qualities into our lives. For many, a path of conscious simplicity involves the integration and balance of both masculine and feminine qualities into a coherent approach to living.

A MEANINGFUL LIVELIHOOD

Those who answered the survey had a variety of work backgrounds. Although for some work seemed to provide little more than a source of income, for most it was a vehicle for participating in the world and a major source of satisfaction in living. Here are some illustrative comments from the letters:

> To me the central reality of voluntary simplicity is the unity and interrelatedness of all aspects of my life. While my work has often been scheduled and distinct from the rest of my life, the more I get into the simpler life, the more I take my work home with me and involve all aspects and experiences of my life in my work. Work is something I do to make a living, but if it's not the kind of work that is also filling my need to feel that I'm contributing in a positive way to the welfare of at least some small part of humanity, I will find some way to make it contributory or find other work. It's like the Eastern perspective that a healthy life is

all a meditation. The more you get into that kind of perspective, the more natural voluntary simplicity is, the easier it is to instinctively do what is going to be good for others and make you happy, and the more consistent and fulfilling your life is going to become.

—*man, thirty-eight, single, small city, West*

It's important for people to realize that this lifestyle places no real boundaries between "work" and "play." "Work" is enjoyed and becomes simply a different activity than "play."

—*woman, nineteen, single, suburb, West*

I live in grateful simplicity. The search for money (for its own sake) interferes in "loving work." For a time it seemed the solution was partial, a compromise between aesthetics and economics—between self-expression and survival. Now it seems that as artificialities are dropped, aesthetics becomes simple and natural—as life energy is released from survival and defense mechanisms, it enters a flow of abundance.

—*man, twenty-nine, single, suburb, Midwest*

In work it takes another kind of energy and self-caring for "sanity" (rather than externally determined "success") to move away from the path that would take me "to the top." I know pretty clearly what the "top" is as defined by my colleagues. What I'm searching for is the "top" as defined by me.

—*woman, twenty-eight, living together, small city, Midwest*

I recognized that the larger American culture did not sustain me with its consumerism and small jobs in large corporations.

—*woman, thirty, living together, rural, West*

Commercialism is making people live on only the periphery of their whole beings. Things don't make you whole and happy, they can divide and disorientate. Putting all of oneself in a task makes you real—whole!

—*woman, sixty-five, single, big city, Midwest*

The attitude toward work that is reflected in these comments seems to be that if we give less than our wholehearted participation to our work, then our sense of connectedness to life itself will be commensurately diminished. Work that is not strongly contributory may yield the income to feed our endless search for gratification, but such work seldom provides us with a sense of genuine contribution and satisfaction. Work that is largely self-serving produces feelings of alienation and unsatisfactoriness. However, when our work is life-serving, then our energy and creativity can flow cleanly and directly through us and into the world without impediment or interruption.

Overall, people viewed work in four primary ways: first, as a means of supporting oneself in activity that is meaningful and materially sustaining; second, as an opportunity to support others by producing goods and services that promote a workable and meaningful world; third, as a context for learning about the nature of life—using work as a medium of personal growth; and fourth, as a direct expression of one's character and talents—as a celebration of one's existence in the world.

Given the drive to find meaningful work coupled with the shortage of such work in today's economy, it is not surprising that many choosing a simpler way of life are involved in starting their own small businesses. Among the various businesses mentioned in the letters were: restaurants, bakeries, bookstores, used-clothing stores, auto-repair shops, bicycle-repair shops, child-care centers, alternative health-care centers, craft shops, grocery stores, alternative schools, and more. This skilled and highly motivated group has ample talent to develop successful and compassionate businesses in communities across the country. Thirty years after the survey, we see that green businesses have been proliferating and their growth skyrocketing, accounting for an estimated $228 billion market in the United States in 2006.[9] Moreover, green business is being promoted in many countries as the remedy for both economic recovery after the meltdown of 2008 and as a healthy response to climate change and the need to reduce greenhouse gas emissions.

SATISFACTIONS AND DISSATISFACTIONS

Given the demands and stresses of this way of life, it would not be voluntarily chosen unless it was satisfying as well. Here is an illustrative sampling of responses to a question asking people to describe their satisfactions and dissatisfactions with living more simply:

> Dissatisfactions . . . are minute, not because they don't exist, but because they are part of the process—not obstacles but bumps on a road that I choose to follow.
>
> —*man, twenty-seven, single, big city, Midwest*

Satisfactions are the fulfillment of the heart. Dissatisfactions are the rumblings of the mind.

—*man, twenty-eight, single, rural, West*

The most satisfying thing is that you can see life right in front of your nose—feel it all around you—running through you and continuing on. It's such a natural occurrence. . . . You gain access to parts of life that are otherwise inaccessible.

—*woman, twenty-three, married, rural, East*

Satisfactions: I am my life. Dissatisfactions: There is always dissatisfaction; it precedes change.

—*man, twenty-seven, married, rural, West*

Satisfaction is internal (not wanting and not rejecting), and I feel it when I am in touch with Reality or God. . . . I always get what I need. This renders my dissatisfactions irrelevant and meaningless (although I forget this often). So I can't bring myself to list my dissatisfactions because they're only a form of unfounded self-pity.

—*man, thirty, married, rural, East*

There are no dissatisfactions, only difficulties, which can and will be overcome.

—*woman, thirty-eight, married, suburb, East*

Satisfactions: Life is a lot simpler—I no longer spend twenty-four hours a month shaving legs and curling hair and God knows how long driving back and forth to Safeway.

Life is infinitely cheaper—releasing money for the real luxuries of life. Dissatisfactions: Outward appearances suggest poverty, and this culture is very discriminatory toward the poor.

—woman, twenty-eight, married, rural, West

The greatest satisfaction derives from increasingly seeing the truth of one of the tenets of simplicity—my needs are always met, although my desires take a beating. . . . I begin to see that my satisfactions and dissatisfactions actually arise more from my attitude than circumstances. This for me is one of the most important aspects of voluntary simplicity . . . the state of consciousness associated with it.

—woman, thirty-two, single, city, West

My life is less cluttered with "things" that control and befuddle me. Dissatisfactions? Only that it's sometimes the harder way to do something. . . . I can't rely on fast food, fast service, fast buying. Everything takes longer—cooking, buying, fixing. But it's worth it—most times.

—woman, twenty-seven, married, rural

The satisfactions are of sharing and caring; of putting forth your best effort regardless of the results; of simply being happy . . . the rewards are immeasurably greater than those of possession or individual accomplishment. After a brief period of remorse for giving up the comfort and recognition that may have been attained, dissatisfaction with life seems to be experienced more infrequently and less intensely than before. As desires become fewer, frustration diminishes. As life becomes less ego-centered, it becomes more enjoyable.

—*man, thirty-one, married, small town, West*

A most satisfying life in that we have a very close family relationship (our children are grown). We see that the children have developed values that are simple and allow for coping flexibly with the changing world. Using our own ideas and hands to make our way in both professions and home life is an exhilarating (and sometimes tiring) way to live.

—*woman, forty-seven, married, suburb, West*

Satisfactions and dissatisfactions: The two sure go together! I can be so elated one day and down in the dumps when the progress is slow the next. Extrinsic joys are: (1) a marriage that works; (2) creating my own joys; (3) making my own music; (4) seeing my spiritual life blossom. Sorrows are just the bottom of the sine wave—when progress is slow.

—*man, twenty-five, married, small town, East*

Satisfaction: Much more flexibility to move, grow, and generally bend with the winds. Continual improvement of my relationship with the universe. . . . In short, after thinking about it a minute my life has only improved since I began to consciously simplify everything. Aside from some complaints about others not understanding my lifestyle (my problem, not the lifestyle's) and a certain difficulty at making money (again, my problem) I have no complaints. . . . I feel more successful, wealthy, and healthy than I ever felt. And I mean wealthy in all senses.

—*man, thirty, single, small town, West*

There are infinite satisfactions—primarily because I have consciously chosen to direct my life toward this lifestyle— I am opening myself to growth, change, freedom of expression, caring for others. The major dissatisfaction—growth is painful!

—*woman, twenty-eight, single, suburb, West*

The pervasive sense of satisfaction expressed in these letters is, I think, a direct result of people's learning to take control of their lives. Taking charge of one's life has many forms: learning basic skills that promote greater self-reliance (gardening, carpentry, repair skills, and so on), choosing work that is contributory and challenging, consuming in ways that respect the rest of life on this planet, and participating in compassionate causes. A growing capacity for self-determining actions contributes to the individual's sense of personal competence, dignity, and self-worth. A positive spiral of learning and growth unfolds as a result of approaching life in this manner. As we become empowered to take charge of our lives, we feel that no one is to blame other than ourselves if our experience of life is not satisfying. And, in our continually opening to and meeting that challenging responsibility, a new sense of freedom, aliveness, and satisfaction naturally emerges.

LESSONS LEARNED FROM THE GRASS ROOTS

There are a number of conclusions that I draw from this survey. First, simpler, more ecofriendly ways of living have taken root not only in a number of developed nations, they are rapidly growing as an integral part of a global, leaderless movement toward sustainable living. Given its emphasis on self-reliance and self-determination, this way of life is tenacious and is blossoming with a diverse garden of expressions that are described in Chapter 1.

Second, the pioneers of sophisticated simplicity demonstrate by their example that we can each take control of our lives. We are not powerless in the face of ecological breakdown. We can create a far more workable and meaningful existence.

Third, an ecological approach to living involves much more than greater material frugality. Like spokes that reach out from the hub of a wheel, this way of life radiates outward from an inner core of experience to touch every facet of life.

Fourth, despite decades of development, lifeways of creative simplicity are still in their springtime of development. Current expressions in the garden of simplicity do not represent the culmination of this way of living; instead, they are its initial blossoming. A vast amount of work and learning yet remains before the potentials of this way of living will be fully apparent.

Fifth, there are no fixed norms that define this approach to living. The worldly expression of an ecological way of life is something that each person must discover in the context of his or her unique life circumstances.

Sixth, these pioneers of Earth-friendly living reveal that "small is beautiful." Small changes that seem inconsequential when viewed in isolation are of revolutionary significance in their cumulative impact. Bit by bit, these many small changes can transform our collective manner of living.

Seventh, those who choose a life of conscious simplicity are reasserting their capacity for citizenship and entrepreneurship. In turn, the traditional polarity of liberal and conservative (concerned primarily with the relative role and power of big government versus big business) is shifting toward another polarity—that of concern for the ability of the individual to determine his or her fate relative to the enormous power of both big government and big business. Traditional political and economic perspectives fail to recognize the most radical change of all in a free-market economy and democratic society: the empowerment of individuals to consciously take charge of their own lives and to begin changing their manner of work, patterns of consumption, forms of governance, modes of communication, and much more.

Eighth, this way of life does not represent a withdrawal from the world. Some may mistake the unwillingness of this forerunner group to participate in the aggressive exploitation of resources, the environment, and other members of the human family as a retreat from the world. Yet, far from withdrawal, a path of conscious simplicity promotes our intimate involvement with life. With conscious and direct involvement comes clarity. With clarity comes insight. With insight comes love. With love comes mutually helpful living. With mutually helpful living a flourishing world civilization is made possible. Rather than abandoning the world, those choosing a life of conscious simplicity are pioneering a new civilizing process.

VALUES OF ENDURING AND BROAD RELEVANCE

The values and views expressed in the Voluntary Simplicity Survey are not a relic of the cultural past; instead, they are alive with relevance. The enduring nature of green values is evident through another in-depth survey taken nearly two decades after the survey described above. In addition, we can see the growing importance of these values in a massive and ongoing world survey. Let's consider each of these.

In 1996, nearly twenty years after I completed the simplicity survey, Linda Breen Pierce gave up a successful career as a lawyer in the fast lane in order to explore a simpler way of living. In the process she conducted in-depth interviews with 211 people in forty U.S. states and eight countries, then reported her findings in her book *Choosing Simplicity*.[10] Her goal was to understand what people's lives were like once they chose a life of "simplicity." The diverse participants represented a wide range of income levels, occupations, and locales, including cities, suburbs, rural areas, and small towns. Each participant answered a detailed survey concerning many aspects of their lives.

Based on her deep inquiry, Linda Breen Pierce came to a number of conclusions about simple living. Perhaps foremost was the understanding that there is no fixed or predetermined way to live simply; Earth-friendly living is something we are each challenged to invent in the unique circumstances of our individual lives. Another conclusion was that many participants see a strong relationship between spirituality and simplicity, so much so that for many it is impossible to separate the two experiences. In turn, the vast majority reported that living simply had enhanced the spiritual aspects of their lives. A further insight from her study was that, for a vast majority of participants, living in a way that respects the Earth is integral to simple living. In these and other ways—the importance of children, family, and community as well as meaningful work—the Pierce Simplicity Study confirms values that have enduring relevance for people.

Turning from the personal to the global, the growing importance of green values is evident in findings from a worldwide survey that has been under way for nearly three decades. The World Values Survey is an ongoing academic project devised by social scientists to assess the state of sociocultural, moral, religious and political values of different cultures around the world. Its results are largely available at www.worldvaluessurvey.com.[11]

This survey represents nearly 70 percent of the world's population and covers the full range of economic and political variation.

Summarizing results from this massive survey, Ronald Inglehart, global coordinator of the survey, concluded that over the past generation a shift in values has been occurring in a cluster of a dozen or so nations, primarily in the United States, Canada, and northern Europe. He calls this change the "postmodern shift." In these societies, emphasis is shifting from economic achievement to postmaterialist values that emphasize individual self-expression, subjective well-being, and quality of life.[12] As well, people in these nations are placing less emphasis on organized religion and more emphasis on discovering their inner sense of meaning and purpose in life.

The values and views found in the original Voluntary Simplicity Survey are consistent with values changes that are measurably under way on a global level. Beneath the surface of modern society, a deep shift has been under way for more than a generation. This shift responds to the excesses of consumer society and reveals a search for greater balance in living.

The Pierce Simplicity Study shows the continuing relevance and richness of simplicity in people's lives. The World Values Survey shows there is a global shift under way as people choose values beyond materialism that are consistent with simpler, greener ways of living. Combining the depth and breadth of these two surveys, we can see that a new constellation of cultural values has been growing beneath the surface of popular culture and has broken into public consciousness as the perfect storm of world problems shatters the myth of perpetual material growth.

three

LIVING VOLUNTARILY

> The little things? The little moments?
> They aren't little.
>
> —*Jon Kabat-Zinn*

SIMPLICITY: VOLUNTARY OR INVOLUNTARY?

It makes an enormous difference whether greater simplicity is voluntarily chosen or involuntarily imposed. For example, consider two people who ride bicycles to work in order to save gasoline.[13] The first person voluntarily chooses to ride a bicycle and derives great satisfaction from the physical exercise, the contact with the outdoors, and the knowledge that he or she is conserving energy. The second person bikes to work because of the force of circumstances—this may be due to the high cost of gasoline or the inability to afford a car. Instead of delighting in the ride, the second individual is filled with resentment with each push of the pedals. This person yearns for the comfort and speed of an automobile and is indifferent to the social benefit derived from the energy savings.

To outward appearance, these two people are engaged in identical activities. Yet the attitudes and experiences of each are quite different. These differences are crucial in determining whether or not bicycling would prove to be a workable and satisfying response, for any given person, to energy shortages. For the first person it would. For the second person it is clearly not a satisfying solution and perhaps not even a workable one (to the extent that he or she, along with many others, tries to circumvent the laws and secure his or her own personal advantage). This example illustrates how important it is that our simplicity be consciously chosen, not externally imposed. *Voluntary* simplicity, then, involves not only what we do (the outer world) but also the intention with which we do it (the inner world).

The simplicity that I focus on is a consciously chosen simplicity. I do not intend to ignore a majority of the human family that lives in involuntary material simplicity—poverty. Rather, I acknowledge that much of the solution to that poverty lies in the voluntary actions of those who live in relative abundance and thereby have a real opportunity to consciously simplify their lives and assist others.

LIVING VOLUNTARILY

To live voluntarily requires not only that we be conscious of the choices before us (the outer world) but also that we be conscious of ourselves as we select among those choices (the inner world). We must be conscious of both the choices and ourselves as the chooser. Put differently, to act voluntarily is to act in a self-determining manner. But who is the "self" making the decisions? If that "self" is both socially and psychologically conditioned into habitual patterns of thought and action, then behavior can hardly be considered voluntary. Therefore, self-realization—the process of realizing who the "self" really is—is crucial to self-determination and voluntary action.

The more precise and sustained our conscious knowing of ourselves, the more voluntary or intentional our participation in life can be. If we are not aware of ourselves as we go through life, then the deliberateness with which we live will be commensurately diminished. The more conscious we are of our passage through life, the more skillfully we can act and the more harmonious our relationships will be: Research has shown that individuals who have higher levels of self-reflection also tend to care more about environmental problems, favor environmental protection over economic growth, and engage in more pro-environmental behavior. Living more consciously fosters an inclusive and compassionate way of living that makes sustainable living possible.

RUNNING ON AUTOMATIC

To appreciate what it means to act voluntarily, it is helpful to acknowledge the extent to which we tend to act involuntarily. We tend to "run on automatic"—act in habitual and preprogrammed ways—to a much greater extent than we commonly recognize. Consider, for example, how we learned to walk as children. At first, walking was an enormous struggle that required all our energy and attention. Within a few months this period of intense struggle passed. As the ability to walk became increasingly automated, we began to focus our attention on other things—reaching, touching, climbing. In the same manner, we have learned and largely automated virtually every facet of our daily lives: walking, driving, reading, working, relating to others, and so on. This habitual patterning of behavior extends into the most intimate details of our lives: the knot we make in tying our shoes, the manner in which we brush our teeth, which leg we put first into a pair of pants, and so on. Not only do automatic patterns of behavior pervade nearly every aspect of our physical existence, they also condition how we think and feel. To be sure, there is a degree of variety in our thinking, feeling, and behaving; yet the variety tends to be predictable, since it is derived largely from preprogrammed and habituated patterns of response to the world. If we do not become conscious of these automated patterns of thinking, feeling, and behaving, then we become, by default, human automatons.

We tend not to notice or appreciate the degree to which we run on automatic—largely because we live in an almost constant state of mental distraction. Our minds are constantly moving about at a lightning-fast pace: thinking about the future, replaying conversations from the past, engaging in inner role-playing, and so on. Without sustained attention it is difficult to appreciate the extent to which we live ensnared in an automated, reflexive, and dreamlike reality that is a subtle and continuously changing blend of fantasy, inner dialogue, memory, planning, and so on. The fact that we spend years acquiring vast amounts of mental *content* does not mean that we are thereby either substantially aware of or in control of our mental *process*. This fact is clearly described by Roger Walsh—a physician, psychiatrist, brain researcher, and lifelong meditator. His vivid description of the nature of thought processes (as revealed in the early stages of meditative practice) is so useful to our discussion that I quote his comments at length:

> *I was forced to recognize that what I had formerly believed to be my rational mind, preoccupied with cognition, planning,*

problem solving, etc., actually comprised a frantic torrent of forceful, demanding, loud, and often unrelated thoughts and fantasies which filled an unbelievable proportion of consciousness even during purposive behavior. The incredible proportion of consciousness which this fantasy world occupied, my powerlessness to remove it for more than a few seconds, and my former state of mindlessness or ignorance of its existence, staggered me. . . . Foremost among the implicit beliefs of orthodox Western psychology is the assumption that man spends most of his time reasoning and problem solving, and that only neurotics and other abnormals spend much time, outside of leisure, in fantasy. However, it is my impression that prolonged self-observation will show that at most times we are living almost in a dream world in which we skillfully and automatically yet unknowingly blend inputs from reality and fantasy in accordance with our needs and defenses. . . . The subtlety, complexity, infinite range and number, and entrapping power of the fantasies which the mind creates seem impossible to comprehend, to differentiate from reality while in them, and even more so to describe to one who has not experienced them.[14]

The crucial importance of penetrating behind our continuous stream of thought (as largely unconscious and lightning-fast flows of inner fantasy-dialogue) is stressed by every major consciousness tradition in the world: Buddhist, Taoist, Hindu, Sufi, Zen, and so on. Western cultures, however, have fostered the understanding that a state of continual mental distraction is the natural order of things. Consequently, by virtue of a largely unconscious social agreement about the nature of our thought processes, we live, individually and collectively, almost totally embedded within our mentally constructed reality. We are so busy creating ever more appealing images or social facades for others to see, and so distracted from the simplicity of our spontaneously arising self, that we do not truly encounter either ourselves or one another. In the process, we lose a large measure of our innate capacity for voluntary, deliberate, intentional action.

So, how are we to penetrate behind automated and habitual patterns of thinking and behaving?

LIVING MORE CONSCIOUSLY

The word "consciousness" literally means "that with which we know." It has also been termed "the knowing faculty." To live more consciously means to be more consciously aware, moment by moment, that we are present in all that we do. When we stand and talk, we know that we are standing and talking. When we sit and eat, we know that we are sitting and eating. When we do the countless things that make up our daily lives, we remember the being that is involved in those activities. We remember ourselves. To "remember" is to make whole. It is the opposite of dismemberment. To live consciously is to move through life with conscious self-remembering.

We are not bound to habitual and preprogrammed ways of perceiving and responding when we are consciously watchful of ourselves in the process of living. Consider several examples. It is difficult to relate to another person solely as the embodiment of a social position or job title when, moment by moment, we are consciously aware of the utter humanness that we both possess—a humanness whose magnificence and mystery dwarfs the seeming importance of status and titles as a basis of relationship. It is difficult to deceive another person when, moment by moment, we are consciously aware of our unfolding experience of deception. When we consciously watch ourselves in the activities of daily life, we cut through confining self-images, social pretenses, and psychological barriers and begin to live more voluntarily.

We all have the ability to consciously know ourselves as we move through life. The capacity to "witness" the unfolding of our lives is not an ability that is remote or hidden from us. To the contrary, it is an experience that is so close, so intimate, and so ordinary that we easily overlook its presence and significance. An old adage states, "It's a rare fish that knows it swims in water." Analogously, the challenge of living voluntarily is not in gaining access to the conscious experiencing of ourselves but rather in consciously recognizing the witnessing experience and then learning the skills of sustaining our opening to that experience.

To clarify the nature of conscious watchfulness, I would like to ask you several questions. Have you been conscious of the fact that you have been sitting here reading this book? Have you been conscious of changes in your bodily sensations, frame of mind, and emotions? Were you totally absorbed in the book until I asked? Or had you unintentionally allowed your thoughts to wander to other concerns? Did you just experience a slight shock of self-recognition when I inquired? What does it feel like to notice yourself reading while you read; to observe yourself eating while you eat; to see yourself watching television while you watch television; to notice yourself driving while you drive; to experience yourself talking while you talk?

Despite the utter simplicity of being consciously watchful of our lives, this is a demanding activity. At first it is a struggle to just occasionally remember ourselves moving through the daily routine. A brief moment of self-remembering is followed by an extended period where we are lost in the flow of thought and the demands of the exterior world. But with practice we find that we can more easily remember ourselves—while walking down the street or while we are at home, at work, or at play. We come to recognize, as direct experience, the nature of "knowing that we know." As our familiarity with this mode of perception increases, we get lost in thought and worldly activities less and less frequently. In turn, we experience our movement through life as more and more intentional and voluntary.

Bringing conscious attention into our daily lives may lack the mystery of searching for enlightenment with an Indian sorcerer and the spiritual glamour of sitting for long months in an Eastern monastic setting, but consciously attending to our daily-life activities is an eminently useful, readily accessible, and powerful tool for enhancing our capacity for voluntary action.

The capacity to move through life with conscious awareness is central to our species identity. We have given ourselves the scientific name *Homo sapiens sapiens*, which means that we are a species that not only "knows" but "knows that it knows." We have identified our core trait as a species—our capacity for reflective consciousness. Living ever more consciously goes to the very heart of our species nature and to our core evolutionary journey as a human community.

EMBEDDED AND SELF-REFLECTIVE CONSCIOUSNESS

Because two different modes of consciousness are crucial to our discussion, I want to define them more carefully. What follows is not an original distinction but an ancient one that has been variously labeled, but similarly described, by many others.[15] The first mode I will call "embedded consciousness" and is our so-called normal or waking consciousness. Embedded consciousness is

characterized by being so absorbed within the stream of inner-fantasy dialogue that little conscious attention can be given to the moment-to-moment experiencing of ourselves. In forgetting ourselves we tend to run on automatic and thereby forfeit our capacity for voluntary action. In the distracted state of embedded consciousness we tend to identify who we are with habitual patterns of behavior, thought, and feeling. We assume this social mask is who we are. Having identified ourselves with this limited and shallow rendering of who we are, we find it difficult to pull away from our masks and freshly experience our identity. Consequently, we feel the need to protect and defend our social facade.

The next step beyond embedded consciousness is what I term "self-reflective consciousness" and provides us with a faithful mirror to impartially witness who we are as we move through our daily lives. Where the distinctive quality of embedded consciousness is self-forgetting (running on automatic), the distinctive quality of self-reflective consciousness is self-remembering (acting in the world intentionally, deliberately, voluntarily). Self-reflective consciousness is not a mechanical watchfulness but a living awareness that changes moment by moment. It means that to varying degrees we are continuously and consciously "tasting" our experience of ourselves. Opening to self-reflective consciousness is marked by the progressive and balanced development of the ability to be simultaneously concentrated (with a precise and delicate attention to the details of life) and mindful (with a panoramic appreciation of the totality of life). Our awareness simultaneously embraces both minute details and larger life circumstances.

Nothing is left out. To make friends with ourselves in this way requires that we accept the totality of ourselves—including

our sensual desires, self-doubts, anger, laziness, restlessness, fears, and so on. We cannot move beyond the habitual pushes and pulls of these forces until we are conscious of their presence in our lives. In turn, to see ourselves in this way calls for great patience, gentleness, and self-forgiveness, as we will notice ourselves thinking and acting in ways that we would like to think we are above or beyond. Yet, to the extent that we are able to see or know our automated patterns, we are then no longer bound by them. We are enabled to act and live voluntarily.

BEYOND REFLECTIVE CONSCIOUSNESS— COMMUNION WITH LIFE

Beyond the duality of a watchful consciousness is the simple unity of resting in the direct experience of the moment and discovering our deep communion with all that exists. The evolution of consciousness does not end with our becoming skilled observers of our life experience. This is the entry into a much larger journey. Self-remembering is the immediately accessible doorway that gradually opens into the farther reaches of conscious knowing. By our being knowingly attentive to the "self" moving through our ordinary, day-to-day life experience, the entire spectrum of conscious evolution unfolds. Just as a giant tree grows from the smallest seedling, so, too, does the seed experience of self-reflective consciousness contain within it the farthest reaches of conscious evolution.

When we tune in to our movement through life, our experience of "self" is gradually though profoundly transformed. As we learn to watch ourselves ever more precisely and intimately, the boundaries between the "self-in-here" and the "world-out-there" begin to dissolve. In the stage beyond self-reflective conscious-

ness, we no longer stand apart from existence as observers; now we are fully immersed within it as conscious participants.

The ability to experience the totality of existence as an unbounded and unbroken whole is not confined to any particular culture, race, or religion. Descriptions of ineffable unity appear throughout recorded history in the writings of every major spiritual tradition in the world: Christianity, Buddhism, Hinduism, Taoism, Judaism, Islam, and all the others.[16] Each tradition records that if we gently though persistently look into our own experience, we will ultimately discover that who "we" are is not different or separate from that which we call God, Cosmic Consciousness, Unbounded Wholeness, the Tao, Nirvana, Brahman, and countless other names for this ultimately unnameable experience. The common experience found at the core of every major spiritual tradition is suggested in the following quotations:

The Kingdom of Heaven is within you.

—words of Jesus

Look within; thou art the Buddha.

—words of Gautama Buddha

Atman [the essence of the individual] and Brahman [the ultimate reality] are one.

—words from the Hindu tradition

He who knows himself knows his Lord.

—words of Muhammad

The experience of unity or wholeness or love lies at the core of every major spiritual tradition. When we come into a conscious relationship with the living universe, we discover common experiences. For example, love is a universal human experience and is not confined to any particular culture, race, or religion. The theologian Paul Tillich described the ultimate nature of love as the experience of life in its actual unity. If love is the experience of life in its actual unity, then consciousness is the vehicle whereby that experience of unity is known. When we become fully conscious of life, we find that it is an unbroken whole. In turn we may describe this experience of wholeness as "love."

THE ENABLING QUALITIES OF
LIVING MORE CONSCIOUSLY

Cultivating our capacity for conscious attention as we move through life is a central project for humanity's future, as it is both empowering and enabling. Being more consciously attentive to our moment-to-moment experience increases our ability to see the world accurately. Given the distracting power of our thoughts (as lightning-fast movements of inner fantasy-dialogue), it is no small task to see things clearly. If we are not paying attention to our flow through life, we will have more accidents along the way, misunderstand others more often, and tend to overlook important things. Conversely, if we are being attentive to ourselves moving through the countless small happenings that comprise our daily lives, then we tend to be more productive; we will listen more carefully and understand more fully, have fewer accidents along the way, and be more present and available in our relationships with others.

Living more consciously has a straightforward and practical

relevance, both for our lives as individuals and as a civilization in a period of stressful transition. As we develop the skills of living more consciously, we are able to see not only our personally constructed reality (as habitual patterns of thought and behavior) but also our socially constructed reality (as equally habitual patterns of thought and behavior that characterize an entire culture). Socially, we are empowered to cut through the political posturing, glib advertisements, and cultural myths that sustain the status quo. In an era dominated by fiercely complex problems of global dimension, the ability to see the world more clearly is essential to the survival and well-being of the human family.

Developing the capacity for self-reflective consciousness also enables us to respond more quickly to subtle feedback that something is amiss. In being more attentive to our situation as a society, we do not have to be shocked or bludgeoned into remedial action by, for example, massive famines or catastrophic environmental accidents. Instead, more subtle signals suffice to indicate that corrective actions are warranted. In the context of an increasingly interdependent world—where the strength of the whole web of social, environmental, and economic relations is increasingly at the mercy of the weakest links—the capacity to respond quickly to subtle warnings that we are getting off a healthy track in our social evolution is indispensable to our long-run survival. As the Internet fosters a new capacity for rapid feedback from citizens and organizations around the world, the human family is developing a level of collective awareness, understanding, and responsiveness to the well-being of the Earth that previously would have been unimaginable.

Living more consciously allows us to respond to situations with greater flexibility and creativity. In seeing our habitual patterns of thought and behavior more clearly, we have greater choice in how we respond. This does not mean that we will always make the "right" choices; rather, with conscious attention our actions and their consequences will become much more visible to us and become a potent source of learning. And with learning comes increasing skillfulness of action.

Reflective consciousness also cultivates feelings of compassionate stewardship toward the rest of life. With conscious attention, we directly sense the subtle though profound connectedness of existence. Awareness of our intimate relationship with the rest of life naturally fosters feelings of compassion. Our range of caring is expanded enormously, and this brings with it a strong feeling of worldly engagement and responsibility.

The ecological crisis we now face has emerged, in no small part, from the gross disparity that exists between our relatively underdeveloped inner faculties and the extremely powerful external technologies at our disposal. With humanity's powers magnified enormously through our technologies, we can do irreparable damage to the planet. The reach of our technological power exceeds the grasp of our inner learning. Unless we expand our interior learning to match our technological advances, we will be destined to act to the detriment of ourselves and the rest of life on the planet. It is vital that we correct the imbalance by developing a level of interior maturation that is at least commensurate with the enormous technological development that has occurred over the last several centuries.

Just as the faculty of the human intellect had to be developed by entire cultures in order to support the emergence of the industrial revolution, we must now develop the faculty of reflective consciousness if we are to make a successful transition to some form of sustainable, global civilization.

THE NATURE OF HUMAN NATURE

Some people believe that "you can't change human nature," and thus see the idea of an evolving human consciousness as no more than unwarranted idealism. Yet, what is human "nature"? The dictionary defines "nature" as the "inherent character or basic constitution of a person or thing—its essence." But does the inherent character and essence of a person ever change? We can gain insight into this key issue by asking an analogous question: Does the inherent character of a seed change when it grows into a tree? Not at all. The potential for becoming a tree was always resident within the seed. When a seed grows into a tree, it represents only a change in the degree to which its potential, always inherent in its original nature, is realized. Similarly, human nature does not change; yet, like the seed with the potential of becoming a tree, human nature is not a static "thing" but a spectrum of potentials. Human beings can grow from a primitive to an enlightened condition without that unfolding representing a change in our basic human nature.

There is, however, a crucial difference between the manners in which the tree and the person realize their innate potentials. For the seed to realize its full expression, it only has to find fertile soil, and the organic cycle of growth unfolds automatically. However, human beings do not develop in an equally automatic manner. For we humans to actualize our potentials, at some point there must be a shift from embedded to self-reflective consciousness (and beyond) if maturation is to continue.

Our culture provides the soil—either moist and fertile or dry and barren—within which we grow. However, the ultimate responsibility for growth, irrespective of cultural setting, remains with the individual. Overall, human nature is not a static condition but an unfolding spectrum of potentials. We can move along that spectrum without changing our basic human nature. That we do progress is vividly illustrated by the fact that humanity has moved from primitive nomad to the edge of global civilization in an instant of geological time. But despite the scope and speed of our evolution, we are far from being fully developed. We are still in the adolescence of our species and have scarcely begun to collectively imagine where our journey could lead in the future.

CONCLUSION

Throughout history, few people have had the opportunity to develop their interior potentials because much of the human journey has been preoccupied with the struggle for survival. The present era of relative abundance—particularly in developed nations—contrasts sharply with the material adversity and poverty of the past. With simplicity, equity, and compassion we can have both freedom from want and the freedom to evolve our interior potentials in cooperation with other members of the human family.

The collective result of a society's cultivating even a modest degree of self-reflective consciousness would be a quantum increase in the effectiveness of self-regulating behavior that is conscious of, and responsive to, the needs of the larger world. The actions of countless individuals would arise from a deeper ground of shared awareness that, in turn, would tend to produce a larger pattern of coherent and harmonious behavior.

Self-reflective consciousness is moving from the status of a spiritual luxury for the few in a more rudimentary and fragmented social setting to that of a social necessity for the many in a highly complex and enormously enlarged social setting. Just as the faculty of the intellect had to be developed by entire cultures to support the emergence of the industrial revolution, so too, I think, must we now begin to cultivate the development of the "knowing faculty," or consciousness, if we are to realize the rapid greening of the world.

four

LIVING SIMPLY

Simplicity is the final achievement . . .
the crowning reward of art.

—*Frédéric Chopin*

THE NATURE OF SIMPLICITY

The dictionary defines "simplicity" as being "direct, clear; free of pretense or dishonesty; free of vanity, ostentation, and undue display; free of secondary complications and distractions." In living more simply we encounter life more directly—in a firsthand and immediate manner. We need little when we are directly in touch with life. It is when we remove ourselves from direct and whole-hearted participation in life that emptiness and boredom creep in. It is then that we begin our search for someone or something that will alleviate our gnawing dissatisfaction. Yet the search is endless in that we are continually led away from ourselves and our experience in the moment.

If you were to choose death as an ally (as a reminder of the preciousness of each moment) and the universe as your home (as a reminder of the awesome dimensions of our existence), then wouldn't a quality of aliveness, immediacy, and poignancy naturally infuse your moment-to-moment living? If you knew that you were going to die within several hours or days, wouldn't the simplest things acquire a luminous and penetrating significance? Wouldn't each moment become precious beyond all previous measure? Wouldn't each flower, each person, each crack in the sidewalk, each tree become a fleeting and never-to-be-repeated miracle? Simplicity of living helps brings this kind of clarity and appreciation into our lives.

An old Eastern saying states, "Simplicity reveals the master." As we gradually master the art of living, a consciously chosen simplicity emerges as the expression of that mastery. Simplicity allows the true character of our lives to show through—as if we were stripping, sanding, and waxing a fine piece of wood that had long been painted over.

SIMPLICITY AS BALANCE OF LIVING

A key figure in the history of simplicity in the West is Richard Gregg. He was a student of Gandhi's teaching and, in 1936, he wrote about a life of "voluntary simplicity." He said that the purpose of life was to create a life of purpose. Gregg saw a life of conscious simplicity and balance as vital in realizing our life purpose because it enables us to avoid needless distractions and busyness. Gregg understood that the nature of one's life purpose—or giving our true gifts to the world—will determine how we arrange our lives. For example, if my true gift is to adopt and raise children, then I may need a large house and car. However, if my true gift is creating art, then I may choose to forgo the house and car and instead travel the world and develop my art. Simplicity is the razor's edge that cuts through the trivial and finds the essential. Simplicity is not about a life of poverty, but a life of purpose. Here is a key passage from Gregg's writing that describes the essence of voluntary simplicity:

> *Voluntary simplicity involves both inner and outer condition. It means singleness of purpose and sincerity and honesty within, as well as avoidance of exterior clutter, of many possessions irrelevant to the chief purpose of life. It means an ordering and guiding of our energy and our desires, a partial restraint in some directions in order to secure greater abundance of life in other directions. It involves a deliberate organization of life for a purpose. Of course, as different people have different purposes in life, what is relevant to the purpose of one person might not be relevant to the purpose of another. . . . The degree of simplification is a matter for each individual to settle for himself.[17]*

There is no special virtue to the phrase "voluntary simplicity"—it is merely a label, and a somewhat awkward one at that. Still, it does acknowledge explicitly that simpler living integrates both inner and outer aspects of life into an organic and purposeful whole.

To live more *voluntarily* is to live more deliberately, intentionally, and purposefully—in short, it is to live more consciously. We cannot be deliberate when we are distracted. We cannot be intentional when we are not paying attention. We cannot be purposeful when we are not being present. Therefore, to act in a voluntary manner is to be aware of ourselves as we move through life. This requires that we pay attention not only to the actions we take in the outer world, but to ourselves acting—our inner world.

To the extent that we do not notice both inner and outer aspects of our passage through life, our capacity for voluntary, deliberate, and purposeful action is diminished.

To live more *simply* is to live more purposefully and with a minimum of needless distraction. The particular expression of simplicity is a personal matter. We each know where our lives are unnecessarily complicated. We are painfully aware of the clutter and pretense that weigh upon us and that make our passage through the world more cumbersome and awkward. To live more simply is to unburden ourselves—to live more lightly, cleanly, aerodynamically. It is to establish a more direct, unpretentious, and unencumbered relationship with all aspects of our lives: the things that we consume, the work that we do, our relationships with others, our connections with nature and the cosmos, and more. Simplicity of living means meeting life face-to-face. It means confronting life clearly, without unnecessary distractions. It means being direct and honest in relationships of all kinds. It means taking life as it is—straight and unadulterated.

When we combine these two concepts for integrating the inner and outer aspects of our lives, we can then say: *Voluntary simplicity is a way of living that is outwardly simple and inwardly rich.* It is a way of being in which our most authentic and alive self is brought into direct and conscious contact with living. This way of life is not a static condition to be achieved, but an ever-changing balance that must be continuously and consciously realized. Simplicity in this sense is not simple. To maintain a skillful balance between the inner and outer aspects of our lives is an enormously challenging and continuously changing process. The objective of the simple life is not to dogmatically live with less but to live with balance in order to realize a life of greater purpose, fulfillment, and satisfaction.

DOES MONEY BUY HAPPINESS?

A key assumption in consumer societies has been the idea that "money buys happiness." Historically, there is a good reason for this assumption—until the last few generations, a majority of people have lived close to subsistence, so an increase in income brought genuine increases in material well-being (e.g., food, shelter, health care) and this has produced more happiness. However, in a number of developed nations, levels of material well-being have moved beyond subsistence to unprecedented abundance. Developed nations have had several generations of unparalleled material prosperity, and a clear understanding is emerging: More money does bring more happiness when we are living on a very low income. However, as a global average, when per capita income reaches the range of $13,000 per year, additional income adds relatively little to our happiness, while other factors such as personal freedom, meaningful work, and social tolerance add much more.[18] Often, a doubling or tripling of income in developed nations has not led to an increase in perceived well-being.

In his book *The High Price of Materialism*, Tim Kasser assembles considerable research showing "the more materialistic values are at the center of our lives, the more our quality of life is diminished."[19] He found that people who placed a relatively high importance on consumer goals such as financial success and material acquisition "reported lower levels of happiness and self-actualization and higher levels of depression, anxiety, narcissism, antisocial behavior, and physical problems such as headaches."[20]

The bottom line is that there is a weak connection between income and happiness once a basic level of economic well-being is reached—roughly $13,000 per year per person. To illustrate this point, the World Values Survey of 2007 revealed that people in Vietnam, with a per capita income of less than $5,000, are just as happy as people in France, with its per capita income of about $22,000. The cattle-herding Masai of Kenya and the Inuit of northern Greenland expressed levels of happiness equal to that of American multimillionaires.

Once a person or family reaches a moderate level of income, here are the factors that research has shown contribute most to happiness:

- **GOOD HEALTH** Physical, emotional, and mental well-being.

- **PERSONAL GROWTH** Opportunities for learning, both inner and outer, and giving creative expression to one's true gifts.

- **STRONG SOCIAL RELATIONSHIPS** Close personal relationships with family, friends, and community in the context of a tolerant and democratic society that values freedom.

- **SERVICE TO OTHERS** Feeling that our lives contribute to the well-being of others.

- **CONNECTION WITH NATURE** Communion with the wildness of nature brings perspective, freshness, and gratitude into our lives.

When we look over this list, it is clear that happiness does not have to cost a lot of money. A tolerant society does not cost a lot in material terms, but the rewards to the social atmosphere in civility, congeniality, and happiness are enormous. Feelings of communion with nature and the cosmos come free with being alive. The quality of relationships with family and community grow from the quality of the time and attention we give to them. Personal growth requires nothing more than paying attention to the experience of moving through life. Feelings of gratitude for life are free.

Happiness is a nonmaterial gift that can spread like a contagion among family, friends, and neighbors—rippling out to touch people who do not even know one another. This is the striking conclusion of a study of more than forty-seven hundred people over a twenty-year period.[21] The study found that one person's happiness can affect another's for as much as a year. Researchers also found that, while unhappiness can spread from person to person like an infection, that emotion appears to be far weaker, and does not spread as far or as powerfully, as happiness. The study also explored the importance of friends and social networks as a source of happiness as compared with the importance of money. The study's coauthor states, "Our work shows that whether a friend's friend is happy has more influence than a $5,000 raise."[22] In the face of economic difficulties, his message is "You still have your friends and family, and these are the people to rely on to be happy." Happiness is a social network phenomenon and can reach up to three degrees of separation (the friend of a friend of a friend), which means that your happiness can involve persons you have not even met.

Happiness is largely a networked social phenomenon once a sustaining level of material well-being is reached. If we worried less about material appearances and thought more about soulful connections with others, we could put our life-energy into creating robust, healthy, and rewarding relationships. The more we learn about the "science of happiness," the more we see that focusing on material acquisition and status is not serving us well and that it would be enormously helpful to redefine progress.

REDEFINING PROGRESS

How can we visualize "progress" in a world that is cultivating lifeways of sophisticated simplicity? The eminent historian Arnold Toynbee invested a lifetime studying the rise and fall of civilizations throughout history, and published numerous volumes. Importantly, Toynbee found a strong connection between simplicity and human progress. Drawing upon his vast knowledge of history, he summarized the essence of a civilization's growth in what he called the Law of Progressive Simplification. He wrote that the progress of a civilization is not accurately measured by its conquest of land and people. Instead, the true measure of growth lies in a civilization's ability to transfer increasing amounts of energy and attention from the material side of life to the nonmaterial side—areas such as personal growth; family relationships; music, theater, and other arts; meditation; community life; personal expression; and democracy.[23] Toynbee invented the word "etherialization" to describe the process whereby, over time, we humans learn to accomplish the same, or even greater, results using progressively less time and energy. "Ephemeralization" is the word that Buckminster Fuller used to describe the process of getting greater material output for less time, weight, and energy invested.

Material ephemeralization is evident in many areas of our lives. For example, computers have evolved from room-size giants to slim laptops or even handheld phones with vastly more computing power. Libraries are being transformed from massive buildings that warehouse millions of books to small computer chips that can store—and intelligently retrieve—an even greater volume of knowledge. Telephone technology has evolved from a heavy network of telephone poles, wires, and transformers to cheap, light, and far more powerful cell phone technologies that use transmitting towers and no longer require cumbersome copper wires strung across the landscape. Automobiles have also ephemeralized as they have advanced from heavy works of iron and steel to an increasingly lighter architecture of high-strength plastic, aluminum, and exotic materials.

Integrating the historical insights of Toynbee and the material insights of Fuller, we can redefine progress as follows: *Progress is a twofold process involving the simultaneous refinement of the material and nonmaterial aspects of life.* With ephemeralization, the material side of life grows lighter, stronger, and more ecofriendly in production, consumption, and recycling. At the same time, the nonmaterial side of life grows in vitality, expressiveness, and insight. Ephemeralization involves the coevolution of inner and outer aspects of life in balance with one another.

The life cycle of an individual provides a useful analogy. From the time that a person is born until his or her late adolescent years, there is usually a tremendous amount of physical growth. Then, in the late teen years, physical growth stabilizes and the person can continue to grow for the rest of his or her lifetime in ways that don't involve growing bigger physically—in physical capacity and skill, in empathy and compassion, in intellectual understanding, and in soulful connection. In a similar way, a portion of our species has experienced a period of extraordinary material growth and is now moving into a stage where further growth could be primarily of a nonphysical nature. In turn, this would liberate resources for those in desperate need and foster a more peaceful world.

Ephemeralism is a coevolutionary approach to living that invites us to continuously balance two aspects of life—maintaining ourselves (creating a workable existence) and surpassing ourselves (creating a meaningful existence). A statement by the philosopher and feminist Simone de Beauvoir helps clarify this: "Life is occupied in both perpetuating itself and in surpassing itself; if all it does is maintain itself, then living is only not dying." On the one hand, if we seek only to maintain ourselves, then no matter how grand our style of living might be, we are doing little more than "only not dying." On the other hand, if we strive only for a meaningful existence without securing the material foundation that supports our lives, then our physical existence is in jeopardy and the opportunity to surpass ourselves becomes little more than a utopian dream.

Ephemeral progress does not turn away from the material side of life; instead, this principle of living calls forth a new partnership in which the material and the nonmaterial aspects of life coevolve and grow in concert with one another. Working together, they can produce ways of living that are materially sustainable, personally rewarding, and culturally rich and engaging. *In place of the failing paradigm of materialism we can choose the promising paradigm of ephemeralism.*

SIMPLICITY AND CONSUMPTION

To live sustainably, it is vital that we each decide how much is "enough." Simplicity is a double-edged sword: Living with either too little or too much will diminish our capacity to realize our potentials. Balance occurs when there is neither material excess nor deficit. To find this in our everyday lives requires that we understand the difference between our needs and wants. "Needs" are those things that are essential to our survival and our growth. "Wants" are those things that are extra—that gratify our psychological desires. For example, we *need* shelter in order to survive; we may *want* a huge house with many extra rooms that are seldom used. We *need* basic medical care; we may *want* cosmetic plastic surgery to disguise the fact that we are getting older. We *need* functional clothing; we may *want* frequent changes in clothing style to reflect the latest fashion. We *need* a nutritious and well-balanced diet; we may *want* to eat at expensive restaurants. We *need* transportation; we may *want* a new Mercedes. Only when we are clear about what we need and what we want can we begin to pare away the excess and find a middle path between extremes. Discovering this balance in everyday life is central to our learning, and no one else can find it for us.

The hallmark of a balanced simplicity is that our lives become clearer, more direct, less pretentious, and less complicated. We are then empowered by our material circumstances rather than enfeebled or distracted by them. Excess in either direction—too much or too little—is complicating. If we are totally absorbed in the struggle for subsistence or, conversely, if we are totally absorbed in the struggle to accumulate, then our capacity to participate wholeheartedly and enthusiastically in life is diminished. Four consumption criteria go to the very heart of the issue of balanced consumption:[24]

- Does what I own or buy promote activity, self-reliance, and involvement, or does it induce passivity and dependence?

- Are my consumption patterns basically satisfying, or do I buy much that serves no real need?

- How tied are my present job and lifestyle to installment payments, maintenance and repair costs, and the expectations of others?

- Do I consider the impact of my consumption patterns on other people and on the Earth?

This compassionate approach to consumption stands in stark contrast to the modern view that consumption is a critical expression of our personal identity. Too often we equate our identity with that which we consume. When we engage in "identity consumption," we become possessed by our possessions, we are consumed by that which we consume. Our identity becomes not a freestanding, authentic expression in the moment, but a material mask that we have constructed so as to present a more appealing image for others to see. The vastness of who we are is then compressed into an ill-fitting and awkward shell that obscures our uniqueness and natural beauty. We begin a never-ending search for a satisfying experience of identity. We look beyond ourselves for the next thing that will make us happy: a new car, a new wardrobe, a new job, a new hairstyle, a new house, and so on. Instead of lasting satisfaction, we find only temporary gratification. After the initial gratification subsides, we must begin again—looking for the next thing that, this time, will bring some measure of enduring satisfaction. But the search is both endless and hopeless, because it is continually directed away from the "self" that is doing the searching.

If we pause in our search and begin to discover that our true identity is much larger than any that can be fashioned through even the most opulent levels of material consumption, then the entire driving force behind our attempts at "identity consumption" is fundamentally transformed. It is transformative to withdraw voluntarily from the preoccupations with the material rat race of accumulation and instead accept our natural experience—unadorned by superfluous goods—as sufficient unto itself. It is a radical simplicity to affirm that our happiness cannot be purchased, no matter how desperately the advertiser may want us to believe the fiction that we will never be happy or adequate without this or that product. It is a radical simplicity when we accept our bodies as they are—when we affirm that each of us is endowed with a dignity, beauty, and character whose natural expression is infinitely more interesting and engaging than any imagined identity we might construct with layers of stylish clothes and cosmetics.

A conscious simplicity, then, is not self-denying but life-affirming. Voluntary simplicity is not an "ascetic simplicity" (of strict austerity); rather, it is an "aesthetic simplicity" where each person considers how his or her level and pattern of consumption can fit with grace and integrity into the practical art of daily living on this planet. Possessions that previously seemed so important and appealing could gradually lose much of their allure. The individual or family admired for a large and luxurious home could find that the mainstream culture increasingly admires those who learn how to combine functional simplicity and beauty in a smaller home. The person who was previously envied for his or her expensive car could find that a growing number of people are uninterested in displays of conspicuous consumption. The person who was previously recognized for always wearing the latest in clothing styles could find that more and more people view high fashion as tasteless ostentation no longer fitting in a world of great human need. All of this does not mean that people would turn away from the material side of life; rather, they would place a premium on living ever more lightly and aesthetically.

SIMPLICITY AND INTERPERSONAL COMMUNICATIONS

The ability to communicate is at the very heart of human life and civilization. If we cannot communicate effectively, civilization itself is threatened. If we apply the principle of simplicity to our communications, they will tend to be more direct, clear, and honest. Here are five areas where simplicity can enhance the quality of communication:

1. Being honest and authentic.

2. Choosing valuable conversations.

3. Valuing the eloquence of silence.

4. Looking with "soft eyes."

5. Respecting physical contact.

First, to communicate more simply means to communicate more honestly—to connect our inner experience with our outer expressions. Integrity, authenticity, and honesty encourage the development of trust. With trust there is a basis for cooperation, even when there remain disagreements. With cooperation there is a foundation for mutually helpful living and mutual respect.

Simplicity of communication, then, is vital for building a sustainable future.

Second, we can bring simplicity into our communication by letting go of wasteful speech and idle gossip. Wasteful speech can assume many forms: distracted chatter about people and places that have little relevance to what is happening in the moment, name-dropping to build social status, using unnecessarily complex or overly coarse language, and so on. When we simplify our communications by eliminating the irrelevant, we infuse what we do communicate with greater importance, dignity, and intention.

Third, simplicity values the place of silence in powerful communication. The revered Indian sage Ramana Maharshi said that silence speaks with "unceasing eloquence."[25] When we appreciate the power and eloquence of silence, our exchanges with others come into sharper focus. The sometimes painful or awkward quality that silence brings in social settings is, I think, a measure of the mismatch between our social facade and our more authentic sense of self. Once we are comfortable in allowing silence its place in communication, there is the opportunity to express ourselves more fully and authentically. The simplicity of silence fosters dignity, depth, and directness in communication.

Fourth, simplicity in communication is also expressed through direct eye contact with others. Because the eyes have been called the seat of the soul, it is not surprising that more direct eye contact with others tends to cultivate more soulful communication. This does not mean engaging others with a tight, hard, and demanding gaze; rather, we can approach others with "soft eyes" that are gentle and accepting. When we directly "see" another in this way, there is often a mutual flash of recognition. The source of that shared awareness resides not in the pigmented portion of the eyes, but within the darkness of the interior center—therein is the place that yields the spark of conscious recognition. It is the dancing and brilliant darkness of the interior eye that reveals that the essences of "self" and "other" arise from the same source. Emerson spoke eloquently of how poverty, riches, status, power, and sex are all forms whose veil yields to our knowing eyes. What is seen goes beyond all of these forms and labels to reveal the very essence of who we are.

Fifth, simplicity can also be expressed in our communications as greater openness to nonsexual physical contact. Hugging and touching that is free from disguised sexual manipulation is a powerful way of more fully and directly communicating with another. Studies have shown that a strong correlation exists between acceptance of physical touching and a tendency toward gentleness. If we are to learn to live together as a global family, then we must learn to touch one another with less physical and psychological violence.

SIMPLICITY AND WORK

Besides powerfully affecting how we consume and communicate, simplicity can transform our approach to work. Our joy and satisfaction with our work is increased greatly when our livelihood makes a genuine contribution both to ourselves and to the human family. It is through our work that we develop many of our life skills, relate with others in shared tasks, and contribute to the larger society. Thomas Aquinas said, "There can be no joy of life without joy of work." Our joy with work can flourish when we move from an intention of "making a killing" for ourselves to that of "earning a living" in a way that contributes to the well-being of all. In sensing and responding to the needs of the world, our work acquires a natural focus and intention that brings clarity and satisfaction into our lives.

Simplicity is also manifest in more human-size places of employment. Many people work within massive bureaucracies: huge corporations, vast government agencies, enormous educational institutions, and sprawling medical complexes. These workplaces have grown so large and so complex that they are virtually incomprehensible, both to those who work within them and to those who are served by them. Not surprisingly, the occupations that often emerge from these massive organizations tend to be routinized, specialized, and stress-producing. Simplicity in this setting implies a change in favor of more human-size workplaces, by redesigning organizations so they are of more comprehensible size and manageable complexity. By consciously creating workplaces that encourage meaningful involvement and personal responsibility, the rampant alienation, boredom, and emptiness of work could be greatly reduced.

Simplicity also finds its expression in more direct and meaningful participation in decisions about work—for example, sharing in decisions about what to produce, being involved in organizing the work process, and taking part in deciding the structure of work arrangements (such as flexible hours, job sharing, job swapping, team assembly, and other innovations).

In looking at three very different areas—consumption, communications, and work—we can see that simplicity has pervasive relevance that can transform our lives. Let's consider some of the more common expressions of simpler ways of living, recognizing that these will evolve as our world changes.

COMMON EXPRESSIONS OF
SIMPLER WAYS OF LIVING

There is no cookbook with clear recipes for the simple life. Richard Gregg, for example, was insistent that "simplicity is a relative matter depending on climate, customs, culture, and the character of the individual."[26] Henry David Thoreau was also clear that no simple formula could define the worldly expression of a simpler life. He said, "I would not have anyone adopt my mode of living on my account. . . . I would have each one be very careful to find out and pursue his own way."[27] Nor did Mahatma Gandhi advocate a blind denial of the material side of life. He said, "As long as you derive inner help and comfort from anything, you should keep it. If you were to give it up in a mood of self-sacrifice or out of a stern sense of duty, you would continue to want it back, and that unsatisifed want would make trouble for you. Only give up a thing when you want some other condition so much that the thing no longer has any attraction for you."[28] Because simplicity has as much to do with each person's purpose in living as it does with his or her standard of living, it follows that there is no single, "right and true" way to live more ecologically and compassionately.

Although there is no dogmatic formula for simpler living, there is a general pattern of behaviors and attitudes that is often associated with this approach to living. After decades of research, I see that many who choose a simpler life:

- Tend to invest the time and energy freed up by simpler living in activities with their partner, children, and friends (walking, making music together, sharing a meal, camping, and so on), or volunteering to help

others, or getting involved in civic affairs to improve
the life of the community.

* Tend to work on developing the full spectrum of their
  potentials: physical (running, biking, hiking, and so
  on), emotional (learning the skills of intimacy and
  sharing feelings in important relationships), mental
  (engaging in lifelong learning by reading, taking
  classes, and the like), and spiritual (learning to move
  through life with a quiet mind and compassionate
  heart).

* Tend to feel an intimate connection with the Earth and
  a reverential concern for nature. When people
  understand that the ecology of the Earth is a part of our
  extended "body," we tend to act in ways that express
  great care for its well-being.

* Tend to feel a compassionate concern for the world's
  poor; a simpler life fosters a sense of kinship with
  people around the world and thus a concern for social
  justice and equity in the use of the world's resources.

* Tend to lower their overall level of personal
  consumption—buy less clothing (paying more attention
  to what is functional, durable, and aesthetic and
  showing less concern with passing fads, fashions, and
  seasonal styles), buy less jewelry and other forms of
  personal ornamentation, buy fewer cosmetic products,
  and observe holidays in a less commercialized manner.

* Tend to alter their patterns of consumption in favor of products that are durable, easy to repair, nonpolluting in their manufacture and use, energy-efficient, functional, and aesthetic.

* Tend to shift their diets away from highly processed foods, meat, and sugar toward foods that are more natural, healthier, simpler, and appropriate for sustaining the inhabitants of a small planet.

* Tend to reduce undue clutter and complexity in their personal lives by giving away or selling those possessions that are seldom used and could be used productively by others (clothing, books, furniture, appliances, tools, and so on).

* Tend to use their consumption politically by boycotting goods and services of companies whose actions or policies they consider unethical.

* Tend to recycle metal, glass, and paper and to cut back on consumption of items that are wasteful of nonrenewable resources.

* Tend to pursue livelihoods that directly contribute to the well-being of the world and enables them to use their creative capacities in ways that are personally fulfilling.

* Tend to develop personal skills that contribute to greater self-reliance and reduce dependence upon

experts to handle life's ordinary demands (for example, basic carpentry, plumbing, appliance repair, gardening, and crafts).

• Tend to prefer smaller-scale, more human-size living and working environments that foster a sense of community, face-to-face contact, and mutual caring.

• Tend to alter male-female roles in favor of nonsexist patterns of relationship.

• Tend to appreciate the simplicity of nonverbal forms of communication—the eloquence of silence, hugging and touching, the language of the eyes.

• Tend to participate in holistic health-care practices that emphasize preventive medicine and the healing powers of the body when assisted by the mind.

• Tend to involve themselves with compassionate causes, such as protecting rain forests and saving animals from extinction, and tend to use nonviolent means in their efforts.

• Tend to change transportation modes in favor of using public transit, carpooling, purchasing smaller and more fuel-efficient autos, living closer to work, riding a bike, and walking.

These are some of the more prominent changes I associate with deep simplicity and green lifeways. Because people around the world are just beginning to invent and create more sustainable and meaningful ways of living, I expect these behaviors to grow and evolve as we move deeper into the perfect storm of a world in systems crisis.

VOLUNTARY SIMPLICITY: AN INTEGRATED PATH FOR LIVING

To live more voluntarily is to live more consciously. To live more consciously is to live in a life-sensing manner. It is to taste our experience of life directly as we move through the world. It is to open ourselves consciously—as fully, patiently, and lovingly as we are able—to the unceasing miracle of our ordinary existence.

To live more simply is to live in harmony with the vast ecology of all life. It is to live with balance—taking no more than we require and, at the same time, giving fully of ourselves. To live with simplicity is by its very nature a life-serving intention. Yet, in serving life, we serve ourselves as well. We are each an inseparable part of the life whose well-being we are serving.

Voluntary simplicity, as a life-sensing and life-serving path, is neither remote nor unapproachable. This way of life is always available to the fortunate minority of the world who live in relative affluence. All that is required is our conscious choosing. This path is no farther from us than we are from ourselves. To discover each of our own unique understandings and expressions of this path does not require us to start from anywhere other than where we already are. This path is not the completion of a journey but its continual beginning anew. Our task is to open freshly to the reality of our situation as it already is and then to respond wholeheartedly to what we experience. The learning that unfolds along the path of our life-sensing and life-serving participation in the world is itself the journey. The path itself is the goal. Ends and means are inseparable.

A self-reinforcing spiral of growth begins to unfold: As we live more consciously, we feel less identified with our material possessions and thereby are enabled to live more simply. As we live more simply and our lives become less filled with unnecessary distractions, we find it easier to bring our undivided attention into our passage through life, and are thereby enabled to live more consciously. Each aspect—living more voluntarily and living more simply—builds upon the other and promotes the progressive refinement of each. Voluntary simplicity fosters:

- A PROGRESSIVE REFINEMENT OF THE SOCIAL AND MATERIAL ASPECTS OF LIFE Learning to touch the Earth ever more lightly with our material demands; learning to serve others ever more responsively with our social institutions; and learning to live our daily lives with ever less complication and clutter.

- A PROGRESSIVE REFINEMENT OF THE SPIRITUAL OR CONSCIOUSNESS ASPECTS OF LIFE Learning the skills of progressively releasing habitual patterns of thinking and behaving that make our passage through life weighty and cloudy rather than light and spacious; learning how to "touch and go"—to not hold on—but to allow each moment to arise with newness and freshness; and learning to be in the world with a quiet mind and open heart.

By simultaneously evolving the material and the consciousness aspects of life in balance with one another—allowing each to inform the other synergistically—we pull ourselves up by our own bootstraps. Gradually the experience of being infuses the process of doing. Life-sensing and life-serving action become one integrated flow of experience. We become whole. Nothing is left out. Nothing is denied. All faculties, all experience, all potentials are available in the moment. And the path ceaselessly unfolds.

five

THE WORLD AT THE
TIPPING POINT

After the final no there comes a yes
And on that yes the future world depends.

—*Wallace Stevens*

Although humanity has always faced challenges, the challenges we face at this time are unique in one respect, and this makes all the difference: The circle has closed. There is nowhere to escape. As a species, we face a systems crisis of global proportions. This crisis also represents a profound opportunity for the human community to pull together in a new way and to respond creatively and cooperatively to our global predicament—a situation unprecedented in human history.

Our first requirement—as individuals, communities, nations, and a species—is to step back and take a very hard look at what is happening with some key trends, such as climate change and running out of cheap oil. Once we have a working grasp of individual trends, we can see how they are interacting with one another in mutually reinforcing ways to produce a world in crisis—in ecological, economical, political, cultural, and other ways. As we come to recognize the magnitude and the urgency of these challenges, we can begin to mobilize ourselves appropriately.

If we misjudge our situation, the results will be catastrophic. There are no "do-overs." We cannot bring extinct species back to life. We cannot refreeze the Arctic and re-create the climate of the past ten thousand years. We cannot refill oil wells that are pumped dry. We cannot replenish ancient aquifers of water that are pumped empty. We cannot take back responsibility for caring for billions of people beyond the carrying capacity of the Earth.

It is very demanding psychologically to consider the breakdown and transformation of civilizations around the world. This is not an abstract process. We are the persons who are living through it. But our anxiety about the changes under way in the world is lessened when we see them as part of a natural and purposeful process—a theme that I will focus on in the following chapter.

In the first chapter, I briefly reviewed some of the powerful trends that are transforming our world. Below I look at these "adversity trends" in more depth as we explore the urgency of change.

GLOBAL ADVERSITY TRENDS

There are dozens of critical trends that we could consider, but I will review seven of them. These are sufficient to give us a clear picture of the magnitude and urgency of the challenges facing us. These trends have been researched for decades, and there is growing scientific agreement concerning each. These trends are:

- Unsustainable population growth.

- Wide and deep poverty.

- Profound climate disruption.

- The end of cheap oil.

- Global water shortages.

- Mass extinction of plant and animal species.

- Unsustainable global footprint.

Unsustainable Population Growth Global population has grown enormously in the past few generations; a few statistics tell the extraordinary story. In the early 1800s, we finally reached a billion persons on the planet. It took another century, until roughly 1930, to reach two billion. Over the next thirty years, we added another billion persons. The three billion humans alive in 1960 doubled to six billion around the turn of the century. As I write this there are nearly seven billion people on the Earth. It is likely we will add another two billion persons to the planet before reaching stabilization at around nine billion people in mid-century. To give a feeling for the magnitude of these numbers, world population growth is roughly seven million persons *each month*—which is the equivalent of adding the population of a city the size of Chicago or Los Angeles to the Earth.

In addition to enormous numbers of humans, we are very rapidly becoming an urbanized species, with nearly all of the growth in world population occurring in the huge megalopolises of the developing world. Throughout human history, the majority of humans lived in rural and small village settings. It was only at the turn of the twenty-first century that we became a predominantly urban species. It is estimated that by the 2020s roughly two-thirds of humanity will live in urban settings. As we become an increasingly urbanized and interconnected world, a new culture and consciousness for humanity is being created.

When most people lived in rural environments, the human family was relatively self-sufficient in providing food. As we become an increasingly urbanized species, we also become less able to provide food for ourselves and more vulnerable to crop failures and famine—a precarious situation in a world that is already experiencing food shortages, as well as climate change and other perils. Overall, both the size and the distribution of global population are magnifying many other difficulties.

Wide and Deep Poverty Although there is a rapidly growing global middle class of roughly two billion persons, there is a much larger portion of humanity (in 2009, roughly five billion persons) living in varying degrees of poverty. For example, in 2009, 75 percent of humans are estimated to live on a *real* income of $4 per day or less. Sixty percent live on a real or effective income of $3 a day or less. The poorest of the poor—subsisting on the equivalent of $1 a day or less—are estimated to number a billion persons. The gap between rich and poor is a vast and growing chasm. In 2001, at the hundredth anniversary of the Nobel Prize, one hundred prize winners came together to issue a "Nobel Warning," whose opening sentence reads: "The most profound danger to world peace in the coming years will stem not from the irrational acts of states or individuals but from the legitimate demands of the world's dispossessed." With the communications revolution, the gap between rich and poor is glaringly obvious.

Even many of the poorest villages have TV sets, cell phones, and computers that give them access to the world. A world that is wired and so profoundly divided cannot work as an integrated and healthy system.

Profound Climate Disruption   Human activities are destabilizing the climate we have enjoyed for the past ten thousand years. Long before glaciers and ice caps melt as a result of global warming, disruption in climate patterns will severely impact the human community. I grew up on a farm and saw the dramatic consequences of modest weather changes. For example, if the rain continued a few weeks longer than usual into the spring, the fields would be too muddy for the farmers to plant their crops. That, in turn, could push harvesting back into the fall. Then, if the frost came earlier than usual in the fall, crops could die or freeze in the ground. Farming and food production are far more vulnerable to climate disruption than many realize.

Looking at the global picture, our world is slipping into a climate crisis much faster than expected. The pace of climate change is exceeding even the most pessimistic forecasts by global climate scientists.[29] The world's foremost authority on climate, scientist James Hansen, states that our situation is "more urgent than had been expected, even by those most attuned." According to Hansen, CO_2's current level of 385 ppm (parts per million)— up from its preindustrial readings of 280 ppm (parts per million)— is far outside its historical levels over millions of years and has already moved into a "dangerous range." Hansen says that if we continue CO_2 buildup into the range of 450 ppm and beyond, as many scientists anticipate, we will create a new planet with a climate unlike anything thus far experienced in human history. The bottom line is that the people of the Earth must take immediate and dramatic action—including transforming consumerist lifestyles—to halt the buildup of CO_2 and then to lower it to 350 ppm or less.[30] If we do not immediately begin to stabilize and then reduce greenhouse gases, we can expect more powerful hurricanes and storms, more intense heat waves and droughts, the spread of diseases, a major shift in growing regions for agriculture, and eventually a large sea level rise that floods coastal regions and produces massive human migrations.

For centuries to come, humanity will be coping with the climate changes set in motion at this time. Global climate destabilization will have such profound consequences that it alone will move humanity onto a new pathway into the future.

The End of Cheap Oil   Petroleum has given humanity a one-time spurt of growth. In rough numbers, we have pumped out roughly half of the oil in the Earth, and this is the "good half" that is relatively easy to reach and get out. The second half gets progressively more costly, as the areas that contain it, such as the deep sea, are difficult to reach and expensive to drill. There is a growing consensus that global oil production has begun to plateau and that we may already have seen the peak in production. At the same time, demand is skyrocketing as countries such as China, India, and Brazil rapidly modernize. China, for example, could use up the world's entire yearly production of oil if it were to modernize and purchase as many cars per capita as does the United States. Given intense demand for oil and a declining supply, we can expect a permanent and growing increase in the cost of oil and oil-based products.[31] This, in turn, will bring profound changes to the structure of the global economy, forcing activities to become much more localized, energy-efficient, and inventive in using renewable sources.

Global Water Shortages   Although we can get by without oil, we cannot live without water, and demand is soaring worldwide. Ironically, although we live on a "water planet," only a small fraction is freshwater that we humans can drink, use for irrigation and growing food, and rely on for manufacturing purposes. More than 97 percent of all the water on the Earth is saltwater. Most freshwater is locked up in the ice sheets of Greenland and the Antarctic. Lack of potable water is blamed for millions of deaths each year from famine, malnutrition, and disease.

Looking ahead, it is estimated that by the 2020s roughly 40 percent of the people on Earth will not be able to get enough water to be self-sufficient in growing their own food.[32] Many will be living in vast urban regions where land and water are scarce. The hardest-hit areas are expected to be northern China, western and southern India, South America, sub-Saharan Africa, and much of Mexico. The United States will also be severely impacted, as, for example, the enormous Ogallala aquifer that sustains the breadbasket of the Midwest is depleted. Looking ahead, the most fierce and protracted conflicts over resources may not be over oil but over freshwater, particularly in these highly water-stressed regions. Even if the energy crisis is solved, the water crisis will grow as global temperatures rise and, for example, cause mountain glaciers to melt, eventually depriving billions of people of a vital source of freshwater.

Massive Extinction of Plant and Animal Species   The overall health of the Earth depends directly upon the well-being of the ecology of animals and plants that live together on the Earth. As we unthinkingly harm that web of life, it can reach a point of no return where it becomes impossible to repair and return to its former status. Currently, the Earth is experiencing one of the largest extinctions of plant and animal species in its four-billion-year history. We are in the midst of the sixth mass extinction, with between one-third and one-half of all plant and animal species at risk of disappearing within a few generations.[33] This is the largest destruction of life on Earth since the dinosaurs and other life-forms were killed by a meteorite roughly 65 million years ago. The fabric of the Earth's ecology is already wounded and torn. We are irretrievably mutilating the biosphere and diminishing the quality of life for countless generations to come.

Unsustainable Global Footprint   Another window onto our collective situation is revealed in humanity's global ecological footprint, which measures the impact of human activities upon the Earth relative to the planet's ability to restore itself. The bottom line is that, since the late 1980s, we have exceeded the Earth's ability to replenish its renewable resources: the fish in the oceans, the fertile farmland, large and healthy forests, and more. To illustrate, in 1961, it is estimated that the human community used roughly half of the world's "biocapacity," or the regenerative capacity of the land, oceans, and air. However, by 1986, human demand began to exceed the regenerative capacity of the Earth. Since then, our relationship with the Earth has become increasingly unbalanced.

Our material demands are overwhelming the Earth's regenerative ability. Within a generation, the human community has become a crushing force impacting every aspect of our planet: overcutting forests, overgrazing pasturelands, overusing farmlands, overfishing oceans, overpolluting the air, and more. We are over the limits in many critical areas, and the growth curve is still headed upward on a path that has already created profound difficulties.

The 2008 "Living Planet Report" by the World Wildlife Fund concluded that humanity is so overusing the Earth's regenerative ability that, if current trends and lifestyles continue, we will require the equivalent of two Earths by the mid–2030s.[34] This is an untenable situation. It will lead to the collapse of ocean fisheries, the ruin of topsoil, the overgrazing and desertification of land, and the depletion of groundwater aquifers—all of which are already occurring.

It has taken the entirety of human history to grow a global economy of the size it reached in 2008. Although our global economy is already exhausting the Earth with unsustainable consumption, if current trends continue, in just two decades the size of the economy will *double*.[35] If the global economy is already overconsuming the Earth, then what will happen when rapidly developing countries with huge populations such as China, India, and Brazil seek to follow their own version of the American, high-consumption lifestyle? Humanity is on a collision course with the Earth, and the time of reckoning has arrived. The reckoning is to find a new way of living that enables us to both maintain ourselves *and* to surpass ourselves as a species.

A PERFECT WORLD STORM

Each of the preceding adversity trends represents an extremely serious challenge. However, when considered together, they represent a supreme test of the evolutionary intelligence and capacity of our species. These trends are of enormous *scale* (often involving millions or billions of people), *complexity* (of bewildering difficulty to comprehend), and *severity* (failure to cope with any one of them will result in monumental human suffering). We confront three types of challenges:

* *Technical* problems (for example, coping with energy and resource shortages).

* *Normative* problems (for example, discovering values beyond materialism that draw people together with a sense of shared purpose).

* *Process* problems (for example, finding ways for millions of citizens to interact with massively complex institutions from the local to the global scale).

These problems comprise a tightly interdependent and *intertwined system of problems* that cannot be dealt with on a one-by-one basis. Instead they require a dramatic shift in our overall pattern of thinking and living.

We cannot say that we were not warned. Over the past several decades, a steady stream of global studies has come to the same conclusion: To avoid an ecological and human disaster, we must make dramatic changes in how we live *now.* Recall the "Warning to Humanity" issued by over sixteen hundred of the world's senior scientists in 1992 alerting the world that we were in danger of "irretrievably mutilating" the biosphere. The United National Global Environmental Outlook Report for 2007, presented as "the final wake-up call to the international community," concluded that the human community is living far beyond its means and inflicting damage on the environment that could soon pass a point of no return. Given these and other warnings, it should come as no surprise that the combined impact of adversity trends confronts the human community with a seemingly insurmountable challenge.

Despite the severity of our physical problems, our deepest challenge is to overcome an invisible crisis: a lack of collective consensus and cohesion around a compelling sense of purpose. What will it take to mobilize humanity's collective efforts in building a green future? Without the beacon of a compelling sense of common purpose, it seems likely that we will withdraw into smaller, more protected worlds. An overriding challenge is to find a new "common sense"—a new sense of reality, human identity, and social purpose that we can hold in common and that respects our radically changing global circumstances. Finding this new common sense in the middle of the turbulence and disarray of the breakdown of civilizations is likely to be a drawn-out, messy, and ambiguous process of social learning. How effectively we use our tools of local-to-global communication to achieve a new consensus will be critical in determining the ultimate outcome.

I do not expect a quick or easy transition through the emerging systems breakdown. Only after people express their anger and sadness over the broken dreams of material prosperity will they turn to the task of building a sustainable economy. Only after people communicate their despair that we may never restore the integrity of the global ecology will we work wholeheartedly for its renewal. Only after people express their unwillingness to make material sacrifices unless their actions are matched fairly by others will a majority of people begin to live in a more ecologically sound manner. Only after people have exhausted the hope that the golden era of growth can somehow be revived will we collectively venture forward. We are moving into a traumatic time of social turmoil that will either transform or devastate the very soul of the species.

There are many reasons to resist change. However, if we (as individuals) fail to get under way with our grassroots adaptation to profoundly changing global circumstances, we (as societies) will be ill prepared when the full force of the global systems crisis hits us. Then the social fabric of nations could be torn apart violently. Democracy could dissolve. Anarchy or, more likely, an authoritarian government could take its place. There are no quick fixes—technological or psychological—to remedy our situation.

It is important not to equate the fulfillment of the potentials of the industrial era with failure. Instead, there has been a spectacular realization of the goals, values, and perspectives of the industrial era. But ironically, these have succeeded all *too* well; we cannot keep going as we have. Now is the time to collectively discover a new beginning—a fresh start with a revitalized sense of purpose as a human community.

There is no one to blame for our time of transition. Who can be blamed when the problems we face are intrinsic to the intertwined structure of life in which we all participate? The views and values that arose in realizing the potential of the industrial era were fitting for that time. Now these same views and values are becoming increasingly ill suited for carrying us into the future.

Although humanity confronts a critical time of transition, the importance of this particular era should not be unduly inflated. In one respect, it is simply another link in a long chain of human evolution. The present era is no more important than any other—failure to forge a strong link anywhere along a chain weakens the entire lineage of development. Yet, in another respect, because the current link in the evolutionary chain can lead to some form of planetary consciousness and consensus, it does seem particularly vital. If we can open to reciprocal learning with other cultures and, through that learning, discover a shared sense of reality, human identity, and social purpose that draws out our enthusiasm for life, then the evolution of humanity will surely lead to unforeseen heights.

In the past the idea of a peaceful and mutually supportive global civilization was viewed as a utopian dream. Now, it is a requirement for continuing human evolution. If we do not rise to this challenge, we will surely unleash upon ourselves the most massive wave of suffering ever experienced in human history. Time has run out. The era of creative adaptation is already upon us.

We can put our predicament into perspective by turning, once again, to the work of the historian Arnold Toynbee. After analyzing the dynamics of development of all the major civilizations throughout history, he concluded that a civilization will begin to disintegrate when it loses its capacity to respond creatively to major challenges. Toynbee also concluded that a failure of creativity often follows a period of great achievement. When we consider industrialized nations and their two centuries of unparalleled material achievement, we should be especially wary of social complacency and diminished creativity. After acquiring a self-image of seemingly invincible mastery, many developed nations find themselves in the demoralizing position of being unable to manage their own affairs, let alone cope creatively with mounting global problems. Redoubling efforts along old lines will fail. It is time to seek out fresh approaches to national and global challenges.

Toynbee was clear in stating that a civilization is not bound to succumb when it enters a stage of breakdown. Particularly in a democratic society, when a civilization fails to employ its inventive capacity and begins to disintegrate, then it is the people themselves who bear ultimate responsibility. Traumatic breakdown is not inevitable, but results from the complacency and inaction of citizens.

TWO OUTCOMES FROM BREAKDOWN

Looking ahead, I see two radically different pathways for humanity. I will symbolize these two futures simply—as either a crash or a bounce. In a *crash*, the biosphere is pushed beyond its ability to support the burden of humanity and suffers crippling devastation. The momentum of historical evolution is dispersed as humanity pulls apart in conflict. In a *bounce*, the same initial conditions prevail but the human community engages in a process of intense communication and reconciliation to build a working consensus around a sustainable pathway into the future. The human family comes together and pulls together during this time of transition, conserving the momentum of historical evolution and building the foundations for a promising future.

Here are two scenarios that suggest how we could evolve into either of these futures.

Scenario I—Evolutionary Crash Because the changes required for sustainability are so fundamental and far-reaching, it is understandable that people may resist and delay until the necessity of making deep changes is unmistakably clear. In his book *Collapse*, Jared Diamond writes that people (both rulers and citizens) often did not see major problems coming (for example, environmental damage, resource depletion, or climate change). Once they did see the problems, they did not recognize their true importance and, as a result, their solutions were weak or inadequate. Being poorly prepared, they were overwhelmed when the full force of change hit. An evolutionary crash could occur if we humans respond to our growing world systems crisis with:

- **DENIAL** Some insist this is not a time of fundamental civilizational transition, that current distress is merely a short-term aberration, and that things will soon return to "normal."

- **HELPLESSNESS** Others feel helpless to make a difference and simultaneously assume that someone else must be in control, so they do not get involved. In feeling powerless to do anything constructive, people adopt a posture of fatalistic resignation toward what appears to be an unstoppable process of disintegration.

- **BLAME** Others assume that some ethnic group, nation, religion, or political group is to blame for these problems, so they invest their creative energy in looking for various scapegoats.

- **ESCAPE** Others acknowledge the seriousness of the situation and look for ways in which they and their friends can escape from a disintegrating situation; some seek to live in self-sufficiency in isolated rural areas, while the wealthy search for secure enclaves in which to "ride out the storm."

In this scenario the mass media, particularly television, is a powerful force in creating the conditions for ecological collapse. By aggressively promoting a consumerist consciousness in order to sell advertising, the mass media dampens public concern and diverts public attention from urgent global challenges. By masking the reality of the world situation, ignoring critical problems, and generating a false sense of normalcy, the mass media retards the process of social learning. As an ill-informed and ill-prepared citizenry loses the race with an intertwined pattern of problems that are growing rapidly in complexity and severity, democratic processes falter and fail.

The grassroots organizations that might have brought an infusion of creativity and innovation into the situation are too few and too weak to make a substantial difference. As a result, people are forced to rely almost exclusively on aging and inflexible bureaucracies that themselves have reached their limits to cope with problems of enormous scale and complexity. With no larger vision beyond that of sheer survival, people and institutions try to hold on to traditional ways of living. With most nations adopting a lifeboat ethic and turning away from responsibilities to the overall biosphere and human family, the world begins a nearly unstoppable slide into calamity.

By responding with too little and too late, the intertwined array of problems (including overpopulation, resource depletion, environmental pollution, species extinction, and climate disruption) exacerbate one another and grow rapidly to devastating proportions. Then, like a rubber band stretched beyond the limits of its elasticity, the global ecology is pushed past its capacity for self-repair and fragments catastrophically. With the global ecology and economy devastated, fierce wars over access to resources (water, farmland, oil, forests, and so on) result, thereby undermining the remaining trust and goodwill essential for working cooperatively to build a sustainable future. Life then degenerates into a survivalist nightmare where the primary concern for most people is the security and survival of their immediate family and friends. The diminished sense of social responsibility of the survivalist orientation accelerates the destruction of the biosphere and produces a downward spiral of war and environmental devastation that feeds upon itself. Within a generation the biosphere becomes so crippled, and the people of the planet so divided by conflict, that a new dark age descends upon the Earth.

Scenario II—Evolutionary Leap Forward   A sustainable future emerges when citizens recognize the absolute necessity of change and use their tools of mass communication to undertake an unprecedented level of dialogue about the most promising pathways ahead. With ongoing, local-to-global communication comes mutual understanding and gradual reconciliation around a shared vision of a sustainable future. With an emerging vision held in common and a commitment to realizing that vision, the human community makes dramatic reductions in military expenditures, begins to heal the global ecology, makes development investments in poorer nations, builds new energy systems, and in many other ways begins to build a promising future.

In building for the long-term future, people see themselves increasingly as part of a tightly interdependent world system rather than as isolated individuals and nations. With a witnessing consciousness or observer's perspective, citizens cultivate the detachment that enables us to stand back, look at the big picture, and make the hard choices and trade-offs that our circumstances demand. With a reflective consciousness, we look at our situation more objectively and, from a global perspective, see how imperative it is to begin the process of healing and reconciliation.

The communications revolution plays a critical role in global consciousness-raising and consensus-building. With the rapid development of sophisticated communication networks, the collective consciousness of humanity awakens decisively. The integration of computers, telephones, television, satellites, and fiber optics into a unified, interactive multimedia system gives the world a powerful voice, and a palpable conscience. The Earth has a new vehicle for its collective thinking and invention that transcends any nation or culture. From this communications revolution comes a trailblazing, new level of human creativity, daring, and action in response to the global ecological crisis.

In multiple ways, mass media and the Internet awaken the collective conversation and creativity of the human community. First, new areas and forms of video programming proliferate as, for example, entire channels are devoted to exploring alternative ways of living and working. Second, the lateral conversation of democracy—citizen-to-citizen dialogue—explodes into an unprecedented level of social communication and consensus-building at every level—local, national, and global. Third, the interaction between an informed citizenry and government leaders also blossoms. Rather than feeling cynical or powerless, citizens feel engaged and responsible for society and its future. As citizens are empowered to cope with mounting crises and to participate in decision making, democratic processes are revitalized. With a free and open exchange of information and visions, and with safeguards to prevent any one group or nation from dominating the conversation of democracy, a foundation for building a sustainable future is firmly established.

A revitalizing society is a decentralizing society, with grass-roots organizations that are numerous enough, have arisen soon enough, and are effective enough to provide a genuine alternative to more centralized bureaucracies. These organizations are taking charge of activities that formerly were handled by state and federal agencies: education, housing, crime prevention, child care, health care, job training, and so on. The strength and resiliency of the social fabric grows as local organizations promote self-help, self-organization, a community spirit, and neighborhood bonding. With control over many of life's basic activities brought back to the local level, a strong foundation is established to compensate for faltering bureaucracies at the state and federal level.

By our breaking the cultural hypnosis of consumerism, developing ecological ways of living, building more conscious and engaged democracies, using the mass media as a potent tool for active social learning, and developing grassroots organizations, a new cultural consensus emerges rapidly. Industrialized nations move beyond the historic agenda of self-serving material progress to a new, life-serving agenda of promoting the well-being of the entire human family. Despite enormous economic, ecological, and social stresses, an overarching vision of a sustainable and satisfying future provides sufficient social glue to hold humanity together while it works through these trying times. A new sense of global community, human dignity, goodwill, and trust grows. Although problems still abound, it is clear that a new springtime of development is emerging.

Two great paths lie before us. How soon might the world hit an evolutionary wall and be forced to choose one path or another?

WHEN WILL WE HIT THE EVOLUTIONARY WALL?

When I was doing research for the first edition of this book in the mid-1970s, key trends such as population, resources, and climate change seemed likely to converge eventually into an intertwined and mutually reinforcing system of trends. At that time, my best estimate was that it would be the decade of the 2020s during which these trends would finally intersect, powerfully amplify one another, and produce an unyielding world crisis—a "supreme test" for humanity. Now, as I write this edition, more than thirty years later, I still think it is likely that in the decade of the 2020s adversity trends will converge into an unyielding systems challenge, producing a decisive tipping point and time of profound choice for humanity.

With economic breakdowns, resource depletion, and climate change already under way, some may argue that the tipping point has already arrived. However, there still seems to be considerable resilience, or "stretch," left in the world system. We have not yet reached the breaking point. Although trends in climate and energy are awakening the people of the Earth to the magnitude of the challenges ahead, they are not yet sufficient to motivate the human community to come together in dialogue around hard choices about our collective approach to material growth. However, by the 2020s, these forces will mature into an unrelenting world systems challenge. The window of opportunity for adaptation is narrowing quickly. Before we hit this evolutionary wall, it is vital that we mobilize our tools of local-to-global communication to meet this challenge. With the push of necessity in alignment with the pull of evolutionary opportunity, it is impossible to imagine the bold and creative actions that could emerge from the collective imagination of humanity as we prepare for our time of profound transition.

CHOOSING OUR FUTURE

Two great alternatives lie ahead—collapse or transformation; ruin or renewal. They overlap at present, but will increasingly diverge as we move into the future. The dimensions of change now occurring are extraordinary. Like giant icebergs breaking loose from ancient moorings and beginning to float free for the first time in thousands of years, entire civilizations are melting, moving, and changing. We have begun a process of transition that extends from the individual to the global scale.

We are not alone in this time of change. Everyone we meet is in some way involved in his or her own personal struggle to respond to our time of transition. Whatever our other differences may be, we are all participants in this historical rite of passage.

As individuals we are not powerless. Opportunities for meaningful and important action are everywhere: in the food we eat, the work we do, the transportation we use, the manner in which we relate to others, the clothing we wear, the learning we acquire, the compassionate causes we support, the level of attention we invest in our moment-to-moment passage through life, and so on. The list is endless, since the stuff of social transformation is identical with the stuff from which our daily lives are constructed.

We are each responsible for the conduct of our lives—and we are each unique. Therefore we are each uniquely responsible for our actions and choices in this pivotal time in human evolution. There is no one who can take our place. We each weave a singular strand in the web of life. No one else can weave that strand for us. What we each contribute is distinct, and what we each withhold is uniquely irreplaceable.

DEEP SIMPLICITY AND
THE HUMAN JOURNEY

From naïve simplicity we arrive at
more profound simplicity.

—*Albert Schweitzer*

In the previous chapter we saw how important it will be to reach into our very soul as a species if we are to find the wisdom to make our way through this pivotal time without irreparably damaging the Earth and ourselves. What insights and stories can guide the world toward a more sustainable and promising future? What is the role of simplicity in enabling those stories to unfold and be expressed? To explore the intimate connection of simplicity with humanity's evolutionary journey, I will explore, first, how simplicity is regarded in the world's wisdom traditions; second, the relevance of simplicity in the emerging paradigm of a living universe; and, third, how simplicity is expressed in evolutionary archetypes that act as beacons for guiding us into the future.

SIMPLICITY IN THE WORLD'S WISDOM TRADITIONS

Simplicity has deep roots in all of the world's wisdom traditions. Although these historical roots are far too extensive to examine here, the following overview shows how simplicity has deep connections with spirituality.[36]

Christian Views   Jesus embodied a life of compassionate simplicity. He taught by work and example that we should not make the acquisition of material possessions our primary aim; instead, we should develop our capacity for loving participation in life. The Bible speaks frequently about the need to find a balance between the material and the spiritual side of life. For example:

- "Give me neither poverty nor wealth." (Proverbs 30:8)

- "Do not store up for yourselves treasure on earth, where it grows rusty and moth-eaten, and thieves break in to steal it. Store up treasure in heaven. . . . For wherever your treasure is, there will your heart be also." (Matthew 6:19–21)

- "Therefore I tell you, do not be anxious about your life, what you shall eat or what you shall drink, nor about your body, what you shall put on. Is not life more than food, and the body more than clothing?" (Matthew 6:25)

- "If a man has enough to live on, and yet when he sees his brother in need shuts up his heart against him, how can it be said that the divine love dwells in him?" (John 3:17)

A common basis for living simply can be found in all the world's spiritual traditions and is expressed in the "golden rule"—the compassionate admonition that we should treat others as we would want ourselves to be treated. The theme of sharing and economic justice seems particularly strong in the Christian tradition.

Eastern Views   Eastern spiritual traditions such as Buddhism, Hinduism, and Taoism have also encouraged a life of material moderation and spiritual abundance. From the Taoist tradition we have this saying from Lao-tzu: "He who knows he has enough is rich."[37]

From the Hindu tradition we have these thoughts from Mahatma Gandhi, the spiritual and political leader who was instrumental in gaining India's independence: "Civilization, in the real sense of the term, consists not in the multiplication, but in the deliberate and voluntary reduction of wants. This alone promotes real happiness and contentment."[38] Gandhi felt that the moderation of our wants increases our capacity to be of service to others. In our being of loving service to others, true civilization emerges. Also found in the Hindu tradition is the idea of *aparigraha*, or "greedlessness" and "nonpossessiveness." As an approach to life, it means to take only what we need and to find satisfaction in that.

Perhaps the most developed expression of a middle way between material excess and deprivation comes from the Buddhist tradition. While Buddhism recognizes that basic material needs must be met in order to realize our potentials, it does not consider our material welfare as an end in itself. Rather, it is a means to the end of awakening to our deeper nature as spiritual beings. The middle way of Buddhism moves between mindless materialism on the one hand and needless poverty on the other. The result is a balanced approach to living that harmonizes inner and outer development.

Early Greek Views   Socrates, Plato, and Aristotle recognized the importance of the "golden mean," or a middle path through life characterized by neither excess nor deficit, but by sufficiency. As in many spiritual traditions, they did not view the material world as primary but as instrumental—as serving our learning about the more expansive world of thought and spirit. Aristotle favored a balanced life that involved moderation on the material side and exertion on the intellectual side. He said that "temperance and courage" were destroyed by either excess or deficiency and could only be preserved by following the golden mean.[39]

Puritan Views   Paradoxically, although the United States is the world's most blatantly consumerist nation, the simple life has strong roots in American history. The early Puritan settlers brought to America their "Puritan ethic," which stressed hard work, temperate living, participation in the life of the community, and a steadfast devotion to things spiritual. Puritans also stressed the golden mean by saying we should not desire more material things than we can use effectively. It is from the New England Puritans that we get the adage, "Use it up, wear it out, make do, or do without." Although the Puritan tradition tended to be hierarchical, elitist, and authoritarian, it also had a compassionate side that encouraged people to use their excess wealth to help the deserving poor. Puritans were not opposed to prosperity itself, but to the greed and selfishness that seemed to accompany excessive abundance.

Quaker Views   The Quakers also had a strong influence on the American character, particularly with their belief that material simplicity was an important aid in evolving toward spiritual perfection. Unlike the Puritans, their strong sense of equality among people fostered religious tolerance. Quakers emphasized the virtues of hard work at one's calling, sobriety, and frugality. Although they thought it only natural to enjoy the fruits of one's labor, they also recognized that our stay on Earth is brief and that people should place much of their love and attention on things eternal.

Transcendentalist Views   Transcendentalist views flourished in the early to mid-1800s in America and are best exemplified by the lives and writings of Ralph Waldo Emerson and Henry David Thoreau. The Transcendentalists believed that a spiritual presence infuses the world. By living simply, we can more easily encounter this miraculous and vital life force. For Emerson the Transcendental path began with self-discovery and then led to "an organic synthesis of that self with the natural world surrounding it."[40] The Transcendentalists had a reverential attitude toward nature and saw the natural world as the doorway to the divine. Nature was seen as the most fitting place for contemplation and for receiving spiritual inspiration. By communing with nature, Emerson felt, people could become "part and parcel with God," thereby realizing the ultimate simplicity of oneness with the divine. Thoreau also viewed simplicity as a means to a higher end. Although he felt that a person "is rich in proportion to the number of things which he can afford to let alone," he was not particularly concerned with the specific manner in which someone lived a simpler life. Instead he was most interested in the rich inner life that could be gained through undistracted contemplation. For both Emerson and Thoreau, simplicity had more to do with one's intentions than with one's possessions.

This brief overview illustrates the long and rich tradition of simplicity of living in human spiritual experience. Historian of the simple life David Shi describes the common denominator among the various approaches to simpler living as the understanding that the making of money and the accumulation of things should not smother the purity of the soul, the life of the mind, the cohesion of the family, or the good of the society. Clearly the simple life is not a new social invention—its value has long been recognized. What is new is the urgent need to respond to the radically changed material and ecological circumstances in which humanity finds itself in the modern world.

DEEP SIMPLICITY IN A LIVING UNIVERSE

According to the Voluntary Simplicity Survey, the most common motivation for moving toward simplicity is to live in a way that integrates and balances the inner or nonmaterial aspects of life with the outer and material aspects. In the survey, some people said they felt uncomfortable calling rich inner experiences "spiritual"; instead, they said, they expressed feelings of deep kinship with nature and satisfaction in their relationships with other people—and wanted to leave explanations at that. However described, it was clear that people were not looking for more stuff but for more happiness, satisfaction . . . and life. In turn, it is *life*—in all its vastness, subtlety, and preciousness—that is the context within which simpler living acquires its compelling meaning and significance. People are reclaiming the subtle aliveness that has been bleached out of everyday existence by the obsessive materialism of the industrial era.

For the past several hundred years, societies in the Western world have been operating under the assumptions of rationalism born of the "Enlightenment Era." Since the late 1700s, scientific belief has been premised on the idea that the universe is a lawful place governed by physical processes. This powerful idea helped free societies from oppressive superstitions such as witchcraft and from authoritarian political systems such as monarchies. It also focused human attention on material things as a source of identity and as a measure of human accomplishment and happiness. With this shift, the Earth came to be regarded as a storehouse of resources for human purposes. Several hundred years later, we are seeing the consequences of this exploitive view in the form of climate change, species extinction, resource depletion, and more. And at the very time the rationalist paradigm has begun to lose its evolutionary relevance and power, a new, and vastly larger, "integral paradigm" has begun to emerge.

Insights from science and spirituality are converging around another perspective that regards the universe as a unique kind of living system. The shift from viewing the universe as nonliving at its foundations to its being recognized as uniquely alive takes us deep into the roots and relevance of simplicity for humanity's future. Let's consider this emerging perceptual paradigm, first from the perspective of the sciences, and second from the perspective of the world's wisdom traditions.

The Science of a Living Universe   For the last several hundred years science has been operating under the assumption that life and consciousness are emergent properties in the universe that appear only when the material world evolves to a high level of biological complexity. Now science is considering the possibility that life and consciousness are fundamental properties of the universe. This is not a new idea. Nearly two thousand years ago, the Greek philosopher Plotinus described the universe as "a single living creature that encompasses all living creatures within it." This ancient insight is being explored anew by modern science.

From the frontiers of science, we are discovering that our universe has a number of key properties of living systems. I realize that some of these are controversial, and I refer the interested reader to my book *The Living Universe*, which explores these in considerable detail.[41] Here I summarize six key attributes of our universe.

- ◆ **COMPLETELY UNIFIED** No longer is the universe regarded as a disconnected collection of planets, stars, and fragments of matter; instead, the powerful tools of science have demonstrated that "nonlocality" exists. Even across vast distances, the universe is fully connected with itself. In the words of physicist David Bohm, the universe is "an undivided wholeness in flowing movement."[42] This does not mean that scientists understand how this connectivity works—only that it is real and that, at a fundamental level, the universe is a completely unified system.

- ◆ **MOSTLY INVISIBLE** Scientists no longer think the visible stars and planets represent all there is in the universe. To their shock, they have recently discovered that the visible universe represents only 4 percent of the total universe—the other 96 percent is invisible! The invisible portion of the universe is comprised of one force that is causing the universe to expand at an increasing rate ("dark energy") and another force causing the universe to contract into clumps of stars and galaxies ("dark matter").

• **IMMENSE BACKGROUND ENERGY** Scientists used to think that empty space was essentially "empty." Now they realize there is an extraordinary amount of background energy permeating the universe, including empty space. Called "zero point energy," a cubic inch of seemingly empty space contains the equivalent of millions of atomic bombs' worth of energy.[43] We are swimming in an ocean of subtle energy of such vast power that it is incomprehensible in human terms.

• **CONTINUOUSLY REGENERATING** The universe is not static, sitting quietly in empty space; instead, the totality of the universe is being regenerated moment by moment—a process requiring a stupendous amount of energy. Despite the appearance of solidity and stability, the universe is a completely dynamic system. In the words of physicist Brian Swimme, "The universe emerges out of an all-nourishing abyss not only fourteen billion years ago but in every moment."[44] At every moment, the universe emerges as a single orchestration—a uni-verse, or single verse, of manifestation. Because nothing is left out of the regeneration of the universe, we are participants in a cosmic-scale process whether we are conscious of it or not.

- **CONSCIOUSNESS AT EVERY SCALE** An ancient and controversial idea is that we can find sentience—some form of consciousness—at every level of the universe. Using sophisticated tools, scientists are now finding a spectrum of consciousness ranging from what might be called primary perception at the atomic and cellular level to a capacity for reflective consciousness at the human level.[45] From the atomic level to the human scale and in between, we find a capacity for reflection and choice that is fitting for that scale.

- **FREEDOM AT THE FOUNDATIONS** We do not live in a machinelike universe where everything is predictable. Instead, at the quantum foundations there is a buzzing world of probabilities and indeterminacy. Uncertainty and freedom are built into the very foundations of material existence. We live in a world of probabilities, not certainties. Freedom and choice are real attributes of the universe and suggest that the universe is a learning system.

When we bring these attributes together, we can then describe the universe as a unique kind of "living system." *The universe is a unified and completely interdependent system that is continuously regenerated by the flow-through of phenomenal amounts of life-energy whose essential nature includes consciousness that supports some freedom of choice at every scale of existence.* This is incredible news. It means we are each completely involved in the regenerative aliveness of the universe. A key function of simplicity becomes apparent—to clear away the clutter, complications, and needless busyness that keep us from connecting with the aliveness that surrounds and sustains us. Simplicity offers a path of intimate encounter with the aliveness permeating the universe.

How Wisdom Traditions View the Universe   When we turn to the "inner sciences," what have sages across cultures and across the centuries discovered with regard to the universe? When men and women from diverse spiritual traditions invest years in deep meditation and contemplation, do they discover the universe to be a place of gray indifference without feeling qualities? Or does the universe reveal itself to be a place of unfathomable mystery, vitality, and wholeness?

The understanding that we live in a living universe is validated and informed by insights from all of the world's major spiritual traditions. Christianity, Islam, Buddhism, Hinduism, indigenous traditions, and more all speak to the idea of a regenerating universe:

> God is creating the entire universe, fully and totally, in this present now. Everything God created . . . God creates now all at once.[46]
>
> —*Meister Eckhart, Christian mystic*

> God keeps a firm hold on the heavens and earth, preventing them from vanishing away.
>
> —*Islamic Koran, 35:41*

> All Hindu religious thought denies that the world of nature stands on its own feet. It is grounded in God; if he were removed it would collapse into nothingness.[47]
>
> —*Huston Smith, scholar of the world's sacred traditions*

My solemn proclamation is that a new universe is created every moment.[48]

—D. T. Suzuki, Zen scholar and teacher

The Tao is the sustaining Life-force and the mother of all things; from it, all things rise and fall without cease.[49]

—Taoist tradition of China

When we harvest the wisdom of human experience, we find remarkably similar descriptions of the universe as a living system. *Christians* and *Jews* affirm that the divine is not separate from this world but continuously creates it anew, so that we live and move and have our being in that divine presence. *Muslims* declare that the entire universe is continually coming into being, that each new moment is a new "occasion" of the universe that went before. *Hindus* proclaim that the entire universe is a single body that is being continually danced into existence by an all-powerful life force, or Brahman. *Buddhists* state that the entire universe arises freshly at each moment in an unceasing flow, with everything depending upon everything else. *Taoists say* the Tao is the "Mother of the Universe" and is the inexhaustible source from which all things emerge without ceasing. *Indigenous* peoples declare that an animating wind or life force blows through all things in the world. With different metaphors a common vision is being told—the universe is a living system that is ceaselessly emerging anew in its totality. We, in turn, are an inseparable part of the ever-arising universe and can learn to come into an ever more conscious relationship with the miracle of existence.

Simplicity in a Living Universe   How does being aware that we live in a living universe transform our lives? If we think the universe is nonliving at its foundations—without meaning or purpose and blind to our existence—it is understandable that to relax is to sink into existential despair. However, if we regard the universe as alive, we can relax into our cosmic home with feelings of curiosity, awe, and love. Here are other ways that it makes an enormous difference whether we see ourselves in a dead or a living universe:

* **CONSUMERISM** Consumerism makes sense in a dead universe. If matter is all there is, then where can I look for happiness? In material things. How do I know my life matters? By how many material things I have accumulated. How should I relate to the world around me? By exploiting that which is dead on behalf of those who are most alive (ourselves). Consumerism and exploitation of the Earth are direct and predictable outcomes from the perceptual paradigm of a nonliving universe. Alternatively, if we regard the universe as a living system, then it is only natural that we look for our happiness in the juice of aliveness—in our relationships with others, with nature, and within ourselves. Living in a living universe, we will naturally seek to reduce needless busyness, complexity, and clutter in the material areas of our lives so that we can have the gift of time and space to engage the nonmaterial areas. As experiential sources of satisfaction become more engaging, consumerism increasingly loses its appeal.

* **IDENTITY** Who are we? What kind of life journey are we on? As we discover that we are intimately connected with a living, regenerative universe, we awaken to a larger sense of identity. Seeing ourselves as part of the seamless fabric of creation awakens our sense of connection with, and compassion for, the totality of life. We no longer see ourselves as isolated beings whose identity stops at the edge of our skin. Instead, we regard ourselves as interconnected beings who are immersed in a vast ocean of aliveness. Our bodies are biodegradable vehicles for acquiring soul-growing experiences.

* **LIFE PURPOSE** A new view of the human journey emerges when we realize that a living universe is also a learning universe. We are learning to live and unfold our potentials in the context of a living universe. We are growing a more intimate relationship with the universe and our soulful nature as we build a sustainable relationship with the Earth and a compassionate relationship with one another.

* **LIFE BALANCE** The world's wisdom traditions have consistently recognized the importance of balance between the material and the nonmaterial aspects of life. Given that the known universe is 96 percent invisible, it is important to not be distracted from the invisible aspects of life by the allure of materialism—the 4 percent of reality. Materialism is the "4 percent solution." A balanced life gives due regard to the invisible intelligence, consciousness, and energy permeating the universe.

* **ETHICS** In seeing that we live within a living universe, we tend to treat everything that exists as uniquely alive and worthy of respect. Every action is felt to have ethical consequences as it reverberates throughout the interwoven field of the living universe. We can tune in to this living field and sense, as a kinesthetic hum, whether our actions are in harmony with the well-being of the world.

The perceptual paradigm of a living universe fosters a new sense of identity, life balance, evolutionary journey, and more. In a mutually supportive process, a lifeway of conscious simplicity both empowers and expresses the perceptual paradigm of a living universe.

SIMPLICITY AND EVOLUTIONARY ARCHETYPES

Another way to see the roots and relevance of simplicity is through the lens of evolutionary archetypes. These are images of the human journey that act as beacons to guide our development as a human community. Evolutionary archetypes are recognized across cultures and provide a common way to visualize who we are and where we are going.

The biggest challenge facing the human community is not the climate crisis, nor the energy crisis, nor the crisis of species extinction; instead, it is a crisis of consensus around a collective vision for a promising future. The human community does not have a compelling story to guide us in responding to a world in systems crisis. Stated differently, we do not yet share in our collective imagination a compelling story of the human journey leading to a sustainable and meaningful future.

If we cannot visualize a future of promise together, then we will pull apart in conflict. As the Bible says, "Where there is no vision, the people perish." Given the severe difficulties facing the world, it is not surprising that most people can readily imagine a future of suffering and ruin, but few can imagine a future of opportunity and renewal. A future of sustainability and promise is still a vague and unformed possibility in our collective imagination. Because we face nearly insurmountable challenges, it will require unprecedented collective vision to transform conflict into collaboration.

Fortunately, several evolutionary archetypes are emerging in the popular imagination and serve as beacons for guiding us into a promising future. We have already talked about the archetype of a "cosmic species"—learning to live in a living universe. This section explores three additional archetypes: the idea that humanity is a maturing species, a heroic species, and a witnessing or observing species. The combination of these four archetypes gives us powerful tools for collectively imagining a positive pathway into the future.

Archetype I—A Maturing Species   One way to regard the human journey is to think of us as maturing over thousands of years. In my experience, people instinctively and intuitively recognize that the human family is growing up. As I've spoken with audiences around the world about humanity's future, I've often begun by asking the following question: "When you look at the overall behavior of the human family, what life-stage do you think we are in? In other words, if you estimate the social average of human behavior around the world, what stage of development best describes the human family: toddler, teenager, adult, or elder?"

When I first began asking this question, I had no idea if people would understand it or how they would respond. To my surprise, around the world, nearly everyone immediately understands this question. With little hesitation, people consistently vote that, as a social average, the overall human family is in its teenage stage of development. To illustrate, in 1999 I posed this question in New Delhi, India, speaking in an auditorium filled with several hundred young schoolteachers who were just graduating from college. There was no confusion about the question. The overwhelming majority voted confidently that the human community is behaving like it is in its adolescent years. I have received similar responses from business leaders in Brazil, spiritual leaders and futurists in Japan, and audiences of all kinds in Canada, Europe, and the United States. All have immediately responded in the same way: approximately three-quarters vote that, as a social average, the human family is in its adolescent stage of development. Many people were quick to point out that this is an average and that therefore some people and cultures are well in advance of this stage.

Around the world, there is clear intuition and understanding that, putting us all together, the human family exhibits many adolescent behaviors. Here are examples often mentioned: Teenagers are *rebellious* and want to prove their independence. Humanity has been rebelling against nature, trying to prove that we are independent from it. Teenagers are *reckless* and tend to live without regard for the consequences of their behavior, often thinking they are immortal. The human family has been recklessly consuming natural resources as if they would last forever. Teenagers are concerned with outer *appearance* and with fitting in materially. Many humans are intensely concerned with how they express their identity and status through material possessions.

Teenagers are drawn toward instant *gratification*. As a species, we are seeking our short-term pleasures and largely ignoring the long-term needs of other species and future generations.

As the perfect storm of a world systems crisis pushes us to move from our adolescence as a species to our early adulthood, our behaviors will shift toward lives of conscious simplicity. As we understand that no outside force is coming to rescue us if we ruin the Earth, we will move from rebelling against nature to cooperating and harmonizing with nature. As we see the big picture of our global footprint, we will move from reckless disregard of our consumption to reflective concern with how we live our lives. As we recognize that short-run gratification is producing long-run ruin, we will extend our horizons of concern to the future well-being of the Earth and all of her creatures. As we appreciate deeper and more fulfilling sources of happiness, we will shift from valuing outward appearances to valuing soulful connection. Overall, a maturing species will move toward a path of conscious simplicity and stewardship of the Earth.

Archetype II—A Heroic Species   A second archetype that can assist humanity in envisioning a positive future is to see ourselves on a heroic journey of collective awakening and development. With this archetype, we can step back from seeing ourselves as "evolutionary villains" who are ruining the Earth and instead regard ourselves as "evolutionary heroes" who are facing a supreme test of our capacities. We are moving through an unprecedented rite of collective passage and confronting the supreme challenge of building a new relationship with the Earth, with one another, and with the living universe that is our home.

The archetype of the hero's journey is widely recognized around the world and is found in stories and myths across history.

The distinguished scholar Joseph Campbell summarized the hero's journey as follows: An adventurer hears a call to discovery and separates from the everyday world, setting out on a search filled with dangers. The hero experiences many difficult challenges and tests, learning much from each. Ultimately, the hero confronts a seemingly insurmountable challenge—a supreme test that cannot be overcome with physical capacities alone. To be successful, the hero must reach beyond his ego and into a spiritual realm where he awakens to a new and more soulful relationship with the Earth, with other people, and with the universe. With this spiritual initiation, the hero then makes a journey of return, bringing these gifts of insight back to the larger community.

Although the popular media often present the hero's journey as a quest for worldly adventure, this is a shallow rendering of this archetype. Instead, the hero's journey is primarily a process of *inner* discovery and personal transformation. In going through a supreme test, the hero surrenders a limited sense of self and awakens to a larger connection with the living universe and the community of all life. In a similar way, we are being called as a species to a vastly larger understanding of who we are and where we are going.

The hero's journey of separation, initiation, and return can be expanded from the scale of an individual to the scope of the entire human community. When we step back and look at the human journey, it is evident that for thousands of years we have been on a path of separation. We have been pulling back from nature and gaining ever more power and control as we've learned to hunt, farm, domesticate animals, build cities, make wars, and transform the planet. Because our power is now so great, we are obliged to be conscious of our intimate connection with nature and to use restraint in the exercise of power. The human community now confronts the supreme test of making a decisive turn to reconnect with nature, one another, and our cosmic home.

How would a heroic species respond to a world in systems crisis? With soulful simplicity, we could turn away from a planet being destroyed by consumerism and focus on what brings the greatest happiness—our relationships with family, community, workmates, and nature. As we make the pivotal turn from separation to cooperation, the human community could heroically self-assemble into a vibrant community that communicates its way into a sustainable and equitable global future.

Archetype III—A Witnessing Species   A third archetype that provides a powerful tool for envisioning a constructive journey for humanity is to see ourselves as a witnessing or observing species. This archetype follows from our scientific name, *Homo sapiens sapiens*. To be "sapient" is to be knowing, so our scientific name means that we are the species that is literally "doubly knowing." What this means is that, where animals "know," we humans have a capacity for "knowing that we know." We have the capacity to bring conscious reflection into choosing our pathway through life.

Despite its utter simplicity, witnessing or watching our lives is an extremely powerful process. It is both healing and transformative to see and accept "what is." All of the world's spiritual traditions as well as psychotherapy are based upon this insight. We grow and evolve by becoming an objective witness or impartial observer of our lives and seeing the reality of our existence. Honest self-reflection cuts through the surface chatter of our lives and reveals the deeper resonance and vulnerability of our soul. As we progressively unfold our capacity to live more consciously, we take soulful responsibility for our lives and relationships.

In a turn from personal consciousness to collective consciousness, within a few decades the combined power of television and the Internet has produced a stunning revolution in our capacity to observe ourselves as a species. With the global communications revolution, the world is becoming transparent to itself. The entire human family is increasingly conscious of the simple fact that we can be conscious of ourselves through the mirror of mass media, the Internet, and other tools of the global nervous system. The world is being enveloped in a growing web of communications and humanity is becoming a collective witness to its own journey.

Although television and the Internet are merging into an integrated media system, it is important to recognize the different strengths of each. Broadcast television can, as it name implies, reach broadly—but its messages are often shallow. The Internet can reach deeply, but its messages are often isolated. By joining deep but disconnected conversations over the Internet with broad but shallow conversations over television, we can transform social communication about our collective future. Neither the Internet nor broadcast television substitutes for the other. However, working together, they can create a broad and deep, resilient and powerful capacity for awakening the collective consciousness and conversation of our species, and for building a working consensus for a sustainable and meaningful future.

Because so much attention is now given to the Internet and television is regarded as "so twentieth century," I want to emphasize how important I think broadcast television is for cultivating a reflective consciousness in modern societies with hundreds of millions of people. I recognize that to suggest television is vital for the functioning of a reflective social consciousness will strike some as an outrageous assertion. Television has been called a "boob tube," a "cultural barbiturate," and a "vast wasteland." How can such a seemingly dysfunctional technology be at the heart of our capacity for social knowing?

Despite the dismal use of this tool of collective communication, the reality is that, even with growing Internet use, television viewing continues to increase.[50] Because most people get most of their news about the world from this source, it means that television creates our shared frame of reference. For all practical purposes, if something does not appear on television, it does not exist in mass social consciousness. Television, then, has become our social witness, our shared vehicle for collective knowing, and it is merging with the Internet to create a globally intelligent, reflective network.

Building a sustainable future will require dramatic changes in the overall levels and patterns of consumption in developed nations. In turn, to change consumption levels and patterns will require a new consciousness and consensus among millions of persons—and this will require dramatic changes in the consumerist messages we broadcast to ourselves through the mass media. In the United States 99 percent of all homes have a TV set, and the average person watches more than four hours of television per day. Despite the growing power of the Internet, a majority of people continue to get most of their news from television. What is more, the average person in the U.S. will see more than thirty-five thousand commercials each year. Television is more powerful than either the schools or the workplace in creating our shared view of reality and social identity, and for promoting lifestyles oriented toward either consumption or conservation.

Despite the fact that in the United States, the airwaves belong to the public and broadcasters are legally obligated to use those airwaves in service of the public interest, the reality is that television broadcasters are aggressively promoting high-consumption lifestyles and are unsympathetic to simpler ways of living. Television stations make their profits by delivering the largest possible audience of potential consumers to corporate advertisers. Mass entertainment is used to capture the attention of a mass audience that is then appealed to by mass advertising in order to promote mass consumption. Although broadcasters have a legal obligation to serve the entire community, in fact the television industry generally ignores the views and values of those who have little to spend (the poor) and those who choose to spend little (the frugal person or family that is more concerned with the quality of being than the quantity of having).

The profound consumerist bias of contemporary television creates an impossible double bind: People use the consumption levels and patterns portrayed in TV advertising to establish their sense of identity and measure their personal well-being while those same consumption patterns are simultaneously devastating the ecological foundations on which our future depends. TV commercials are far more than a pitch for a particular product— they are also advertisements for the attitudes, values, and life-styles that surround consumption of that product. The clothing, cars, settings, and other elements that create the context for an advertisement send strong, implicit messages about the standards of living and patterns of behavior that are either the norm or the ideal for society. More sustainable patterns of living and consuming seldom appear on television, as these themes would threaten the legitimacy and potency of the television-induced cultural hypnosis generated by mass entertainment, mass advertising, and mass consumption. By default, societies are left with programming and advertising that selectively portray and powerfully reinforce a materialistic orientation toward life. *Because television's being programmed to achieve commercial success, the mind-set of entire nations is being programmed for ecological failure.*

The most precious resource of a civilization—the shared consciousness of its citizenry—is literally being prostituted and sold to the highest corporate bidders. Each time we watch commercial television, we are putting our collective values, attitudes, and priorities up for sale. The pervasive commercialization of television, and thus society, represents far more than an offense to "good taste"—it is crippling our capacity to comprehend and respond to mounting world challenges. Commercialization is distorting and undermining the very foundation of civilization— the view of reality and social identity that we hold in common. Television advertising takes exceedingly trivial concerns (such as which deodorant, shampoo, or denture adhesive to use) and blows them up into issues of seemingly enormous importance for our lives. Concerns that are utterly insignificant relative to the task of making it through this time of profound ecological and social transition are given vastly inflated significance and then force-fed into our collective consciousness. *To break the cultural hypnosis of consumerism, we must begin by breaking the corporate stranglehold on broadcast television.*

We are a witnessing species and have powerful technologies at our disposal that can transform how we see ourselves and the human journey. Yet, the window onto the world offered by mainstream television is so obsessed with consumption and so disconnected from a world moving into a whole-systems crisis that it is retarding rather than enabling human adaptation. The point is not to condemn all advertising and entertainment; rather, it is to acknowledge that we are a witnessing species and we require a new diet of images and information if we are to move swiftly toward a sustainable future. By mobilizing the observing capacity of our powerful media, we can bear witness to urgent concerns such as climate change, peak oil, water vulnerability disparities between rich and poor, and species extinction. We can also call forth fresh visions of the future and bring them into our collective consciousness. Our media environment—dominated by television and the Internet—both reflect and constrain the workings of our collective mind. As we collectively think, so will we go; therefore, as the media goes, so goes our future.

We have considered four powerful archetypes in an effort to tell a new story about the human journey. Through the lens of these archetypes, we see that: We are a *maturing* species that is entering a rite of passage that can take us from our adolescence to our adulthood. Living simply and sustainably is a direct expression of our growing maturity. We are a *heroic* species that has been progressively separating ourselves from nature and becoming ever more differentiated and powerful; now we are beginning a journey of return to the living universe that is our home. Our journey of return is one of communion and compassion—a new caring for life is expressed in our choice of simplicity "so that others may simply live." We are a *witnessing* species that has been moving through history half awake, not fully utilizing our unique capacity for being conscious; now—aided by the global communications revolution—we are becoming more fully awake and choosing our pathway ahead more deliberately. We are a *cosmic* species, as we have the ability to be conscious of our connection with and participation in the cosmos.

When we combine these four archetypes, we can summarize the purpose and promise of the human journey as follows: *Humanity is on a heroic journey of awakening into the stunning reality that we are beings of cosmic connection and participation who are learning to live within a living universe.* Our potentials as a species are as magnificent and mysterious as the living universe that permeates and sustains us.

SIMPLICITY AND THE HUMAN JOURNEY

A lifeway of conscious simplicity is integral to our journey, as it both enables and expresses the next stages in our collective journey: maturing into our early adulthood as a species; making the heroic journey from individuation back to communion; growing our capacity for a witnessing consciousness as a species; and learning to live ever more fully and creatively in our living universe.

seven

LIVING IN A GREEN WORLD

> Practically speaking, a life that is vowed to simplicity, appropriate boldness, good humor, gratitude, unstinting work and play, and lots of walking brings us close to the actual existing world and its wholeness.
>
> —*Gary Snyder*

What would life be like in a greening world? Without romanticizing, how would life be different if a majority of people were to choose a path of conscious simplicity? Looking at the inevitable shocks and hardships that lie ahead, green approaches to living must be hardy, resilient, and tenacious.

The world confronts not only a number of individual crises, but additionally a megacrisis as individual trends mutually reinforce one another and create a larger world systems emergency. Reviewing briefly, here are eight major emergencies now facing the human family:

1. *Climate emergency:* We are destabilizing the climate, and this is producing crop failures, famine, food riots, and social breakdowns.

2. *Energy emergency*: The world is running out of cheap oil at the very time that energy demand is sky-rocketing. Development of renewable sources of energy is lagging far behind, creating a difficult transition ahead.

3. *Water emergency*: Freshwater is in short supply in a number of parts of the world, and as aquifers are pumped dry and global warming melts snow and glacier ice, water shortages will soon reach crisis proportions.

4. *Sustainability emergency*: The demand for resources is growing rapidly and already exceeds the regenerative capacity of the Earth.

5. *Poverty emergency*: Three-quarters of the human family lives on a real income of $4 per day or less at the same time that we live in an increasingly wired and transparent world where differences are glaringly obvious.

6. *Financial emergency*: The world financial system is not serving the well-being of the people but instead the profits of powerful financial institutions, and the result is global financial chaos and breakdowns.

7. *Extinction emergency*: Between one-third and one-half of all plant and animal species are currently threatened with extinction as the web of life on the Earth is being devastated by humanity's unthinking growth.

8. *Weapons of mass destruction emergency:* At the very time the world is becoming much more interdependent and vulnerable to disruption, we are experiencing the spread of terrifying weapons of mass destruction (biological, chemical, and nuclear).

Any one of these eight emergencies could bring the human enterprise to grief. When they converge at the same time, they constitute far more than an emergency—they represent an unequivocal catastrophe. In this generation we will either devastate or transform the human journey.

We have no place to escape from our global predicament. We are being called to rise above the divisions and differences of the past. Working together we can build a promising future for our children and grandchildren. Choosing sustainability will require billions of persons to come together with a level of social consciousness, consensus, and commitment that has never before existed in human history. Were it not for the push of the most urgent and unyielding necessity, few people would choose a path as bold, inclusive, and collaborative as is demanded by our times. Our choice is ruin or responsibility. If we are to take responsibility for our growing world systems crisis and build a workable and meaningful human civilization, then major changes are required, both outer and inner.

OUTER EXPRESSIONS OF GREEN LIVING

How would everyday life be different in a green world? The most noticeable changes in the everyday life would concern the basics—the community in which we live, the work we do, the housing in which we live, the transportation we use, the food we eat, the clothes we wear, the education we acquire, and so on. Let's bring these themes together in a brief scenario:

In a world systems crisis, the most sustainable places to live will be towns and small cities that are nested within a larger agricultural area. As the cost of energy becomes increasingly expensive, the least well-adapted places will likely be high-rise cities, which are voracious consumers of energy and generally far from agricultural resources. Also poorly adapted are sprawling suburbs that are far from any significant economic core and food sources. Particularly in the United States, we can see the enormous misallocation of resources represented by urban sprawl, in which more than half of the population lives in suburbia. In the years ahead, we will see waves of migration toward towns and regions more favored for sustainability by weather and local resources.

Without cheap oil to fuel an extravagant global economy, growth must become much more selective and differentiated. Some sectors of the economy are contracting (especially those that are wasteful of energy and oriented toward conspicuous consumption), while others are expanding (for example, Internet-based businesses, intensive organic agriculture, retrofitting homes and cities for energy efficiency, and education for skills in sustainability). Markets are becoming increasingly decentralized and localized in order to reduce transportation costs in this energy-expensive setting. Food is also impacted by energy costs and is increasingly locally grown by smaller organic farmers. As people rely increasingly upon locally grown food, diets adapt to foods that are in season. Buying goods and services locally fosters a rebirth of small businesses and entrepreneurial activity, and overall, small businesses that are well adapted to local conditions and local needs are prospering. New types of markets and marketplaces are growing—flea markets, community markets, and extensive bartering networks. Alternative or local currencies are flourishing, promoting engagement with the local economy.

The nature of work is also changing as a growing proportion of people seek to integrate their work lives and home lives into a community experience. Instead of a single job, many work at several jobs requiring different skills. People are compensated with a mixture of barter, local currency, and national currency. The shift to green living has opened up an entire industry in retrofitting buildings, developing green businesses of a village scale, developing locally sustainable agriculture, and so on.

Energy conservation and an ecological consciousness is widespread and visible everywhere. Solar panels cover many roofs. An edible landscape is growing, with intensive gardens in vacant city lots and on rooftops. In the streets are fuel-efficient autos and trucks.

Consumption patterns are shifting in favor of products that are functional, durable, energy-efficient, nonpolluting, easily repairable, healthy, and produced by ethical firms. A new consumer consciousness is growing as people boycott goods and services sold by firms whose policies are considered unethical or unsound with regard to their treatment of the environment, workers, overseas investment policy, and so on.

There is an explosion in websites, blogs, books, and classes dealing with every aspect of sustainable living—from ecovillage living to home retrofitting, seasonal cooking, and gardening. Everywhere there is a feeling of rebirth in personal competence and craftsmanship.

The foregoing only begins to suggest the practical, down-to-earth adaptations that would characterize a revitalizing civilization. In countless small ways, nearly every aspect of life will be adapted to the reality of the global ecology. Voluntary simplicity does *not* mean a return to a more primitive past but rather a movement ahead to a more sophisticated, compassionate, and collaborative future.

INNER EXPRESSIONS OF GREEN LIVING

We may think we are powerless to create a promising future. In reality we have at least six great powers that we can mobilize in order to transform our time of adversity into opportunity:

1. *Perception:* We can choose to explore the universe around us and the permeating aliveness recognized by the world's wisdom traditions.

2. *Choice:* We can choose whether or not to live in a manner that is sustainable.

3. *Community:* We can choose how we live in our local neighborhoods and communities and the extent to which we work to make them sustainable, safe, and friendly.

4. *Communication:* We can choose how we use our capacity for communication, from personal conversations to global connections via television and the Internet.

5. *Democracy:* We can choose how we inform ourselves as citizens, build a working consensus for the future, and engage our leaders in working for change.

6. *Love:* We can choose how we relate to one another as global villagers and the degree to which we live in accord with the commonly recognized golden rule.

Let's consider each of these. Because the first two empowerments have already been considered at length, I mention them only briefly:

Power of Perception   Our direct experience of the subtle aliveness infusing the universe is transformative. When we relax into our direct experience, we rest within the ecology of conscious aliveness, and this expands our vision of the human journey. We intuitively recognize that we are learning to live in a living universe.

As the paradigm of a living universe infuses the view of reality that we hold in common, we will feel more at home with the world around us, knowing it has been welcoming our evolution and awakening for billions of years. As we grow into the aliveness, material riches lose their former appeal compared with the rewards of experiential riches of spending time in nature's wildness, engaging in meaningful relationships, and doing work that contributes to the well-being of life. In approaching life this way, our soulful potentials unfold as we seek harmonious relationships with one another, a sustainable relationship with the Earth, and a sacred relationship with the living universe.

Power of Choice   A core theme throughout this book is that we are not powerless in creating a sustainable future. Our personal lives are filled with choices: the food we eat, the home in which we live, the work we do, the clothes we wear, the transportation we use, and much more. We can voluntarily orient toward sustainability. We can choose Earth-friendly, or green, living. We can choose to live moderately on the material side of life and to put increasing energy and attention into the nonmaterial side of life. We can explore the "garden of simplicity" and discover expressions that fit our own lives. We can choose reconciliation instead of conflict. We can choose, in our personal lives, to help heal the wounds of history and to create a strong bond among the human community as a foundation for the future. We can choose a higher path for humanity that honors the deepest wisdom of the world spiritual traditions and the sciences. We can see ourselves on a sacred journey of awakening to our participation in the unfolding of a living universe. Simplicity of living both enables this journey and is an expression of its fulfillment.

Power of Community   We have the power to create sustainable and satisfying forms of community. Unfortunately, our current scales of urban living often do not serve living in community. At the scale of a household, we are so small and our energies so disconnected that there is little we can do. At the scale of a whole city, cooperation becomes so cumbersome and massive as to be unworkable. However, at the scale of the neighborhood or a small village with a hundred or so persons, we find a size that is both congenial and workable. The village scale is small enough to support strong personal relationships and yet large enough to also support community activities. It is in this in-between zone that a new village movement can flourish and people will find a scale of life that is both workable and meaningful.

In the future, many neighborhoods with single-family dwellings will be configured into unique clusters of small village communities. Modern cohousing communities or ecovillages are showing the way to make this work. To illustrate, my wife Coleen and I lived in a cohousing community in Northern California for a year and a half. The three core organizing principles for the community are simplicity, family, and ecology. With seventy people (fifty adults and twenty children), this was a scale of living that was small enough to create a genuine feeling of community and large enough to use our size to advantage. We lived in a newly constructed community consisting of thirty units in two-story flats and town houses clustered in rows to establish a common green area on the interior and parking on the exterior. The common house is used as a dining area but is regularly transformed into a dance floor, meeting room, or whatever else it needs to be. The common house also includes two guest rooms, an informal lending library, and a playroom for small kids on rainy or cold days.

As a community, we would eat together three evenings each week and often join up for a brunch on weekends. Each person is expected to participate in a three-person cooking crew once a month, preparing food and cleaning up for roughly fifty persons. People are also expected to participate in work crews to perform functions like landscaping, conflict resolution, or kitchen maintenance. Every other week there are meetings to run the workings of the community. Happily, these are run efficiently and expertly, attendance is high, and much is accomplished. This cohousing community also has a half-dozen office and commercial spaces connected with it (a coffee shop, a green consulting business, a copy shop, etc.), so it is both a housing entity and a commercial enterprise.

Beyond the formal activities of operating a cohousing community were the informal ones that brought us together in strong relationships. We easily and quickly organized diverse activities ranging from fundraisers (such as a brunch for tsunami disaster relief) to classes (such as yoga and Cajun dancing) to community celebrations and events. Again and again, we saw diverse gatherings and initiatives emerge from the combined strengths and diverse talents of the community.

Envisioning a sustainable future, families will live in an "ecohome" that is nested within an "ecovillage," that, in turn, is nested within an "ecocity," and so on up to the scale of the bioregion, nation, and world. Each ecovillage of one or two hundred persons could have a distinct character, architecture, and local economy. Common to many of these villages would be a child-care facility and play area; a common house for community meetings, celebrations, and regular meals together; an organic community garden; a recycling and composting area; an open space; and a crafts and shop area. As well, each would offer a variety of types of work to the local economy, such as child care, aging care, organic gardening, green building, conflict resolution, and other skills that provide fulfilling employment for many. These microcommunities or new villages could craft unique expressions of sustainability while creating meaningful work, raising healthy children, celebrating life in community with others, and living in a way that seeks to honor the Earth and future generations.

A new village movement could transform urban life around the world. Drawing inspiration from cohousing and ecovillages, diverse communities would flower, replacing the alienating landscape of today's massive cities and homogeneous suburbs. Ecovillages provide a practical scale and resilient foundation for a sustainable future and could become important islands of security, camaraderie, learning, and innovation in a world of sweeping change. These human-size living environments will foster diverse experiments in cooperative living arrangements that touch the Earth lightly and that are uniquely adapted to each locale.

Although ecovillages are designed for sustainable living, there is not the time to retrofit and rebuild our existing urban infrastructure around this approach to living before we encounter the world systems crisis. Climate disruption, energy shortages, and other critical trends will overtake us long before we can make a sweeping overhaul in the design and functioning of cities and towns that have been a century or more in the making. We can regard ecovillages as greenhouses of human invention, learn from their experiments, and adapt their designs and principles for successful living.

Without the time to retrofit cities into well-designed "green villages," we must make the most of the urban infrastructure that already exists. Creatively adapting ourselves to this new world of climate change and energy shortages is producing a wave of green innovations for local living—technical, social, architectural, and more. A self-organizing grassroots movement is now emerging around the world to facilitate the transition to urban designs for sustainable living.

"Transition towns" is a name for the process of adapting to the world-changing trends of peak oil and climate change. The "town" can range in size from a small community to a large urban system. The idea for "transition towns" emerged in 2005 in the UK as Rob Hopkins's student project, and rapidly went viral—spreading around the world in less than three years, with grassroots initiatives springing up in hundreds of diverse communities.

A transition town begins with a small group of motivated persons who share a common concern: How can this community respond to both the challenges and the opportunities of peak oil and climate change? Organizers of this movement say that, because of climate change, it is *essential* that we shift to a renewable energy economy and society. Because of peak oil, it is *inevitable* that we will make this shift. Because the future that can emerge is more habitable and sustainable, it is *desirable*. Understandably, because it is essential, inevitable, and desirable, this is a movement under way around the world.

Organizers of transition towns are seeking to prepare communities for the inevitable shocks to the economy and society that will come from oil shortages and climate disruption. A transition town seeks to build resilience in advance and create the ability to bounce back after predictable shocks to the system. The Transition Handbook shows how organizers see a different and more favorable future ahead, one with:

> . . . the potential for an economic, cultural, and social renaissance the likes of which we have never seen. We will see a flourishing of local businesses, local skills and solutions, and a flowering of ingenuity and creativity. It is a transition in which we will inevitably grow, and in which our evolution is a precondition for progress. Emerging at the other end, we will not be the same as we were: We will have become more humble, more connected to the natural world, fitter, leaner, more skilled, and ultimately wiser.[51]

Pushed by necessity and pulled by opportunity, a self-organizing transition town movement is under way around the world. Communities are looking ahead fifteen to twenty years and thinking how they can build resilience. Despair is replaced by excitement as people come together to meet foreseeable challenges. Instead of trusting state and national governments to solve things, people are drawing upon the unique strengths and resources of the local community to build a sustainable future.

Not that the organizers of this movement think they have all the answers. Here is a refreshing admission from the organizers:

> We truly don't know if this will work. Transition is a social experiment on a massive scale. What we are convinced of is this: If we wait for the governments, it'll be too little, too late; if we act as individuals, it will be too little; but if we act as communities, it might just be enough, just in time.

The honesty, authenticity, and maturity of this approach is challenging people from all walks of life to get involved in adapting to the new world. At the foundation of this process is the individual or family choosing greener, simpler ways of living at the personal level. In turn, this provides the ground for building resilient econeighborhoods and ecovillages on the scale of a few hundred people who work and live in a purposeful manner. Clusters of retrofitted neighborhoods then create a resilient base for an entire town that is in transition to sustainability. The efforts of the town are, in turn, supported by initiatives at the state, federal, and global levels to build a sustainable future. Ultimately, no one is in charge—the overall system is being designed by everyone, working at every level. With everyone taking their share of responsibility, the sum total of our collective actions have the potential to rapidly build a resilient and sustainable world system.

The question of questions is whether this transition can take place fast enough. We are in a race with ruin. Recognizing instinctively that our survival is at stake, the human community may reach beneath the surface chaos of our times and discover a deeper wisdom that, although diversely expressed, is collectively recognized and orients us toward a more promising future. Critical to discovering our collective wisdom and a new common sense for the future is making mature use of our system of global communications.

Power of Communication   Our ability to communicate has enabled humans to evolve from awakening hunter-gatherers to a developing planetary civilization. If we are to make it through this rite of passage as a species and become a mature planetary civilization, our ability to communicate will have to play a decisive role. "Communicate or perish" is fitting wisdom for our times.

When people ask me, "What can I do?" I often reply that one of the most powerful things we can do is to start talking with other people about our personal hopes and fears for the future. Our personal conversations with others create the fabric of our culture and the foundations of our democracy. Our conversations extend into the world and help weave together a tapestry of shared understanding and consensus for our pathway ahead. There are innumerable places where we can engage in powerful conversations about our common future—in boardrooms, classrooms, living rooms, houses of worship, and more. If we are genuinely willing to look for common ground rather than just preach to others, we can strike up mutually helpful conversations about the challenges and the opportunities that lie ahead.

To be successful, the scale of humanity's conversations must ultimately match the scale of the challenges. Because many of the crises facing humanity are of global scale, it is vital that we have conversations at this larger scale as well. Fortunately, we have the tools to realize this unprecedented level of communication.

At the very time the entire Earth is under assault and we require a capacity for local-to-global communication, the revolution we need is well under way. With the Internet weaving the world into a unified whole and television providing us with a visual literacy for connection, the capacity for a new culture and consciousness is entering the Earth. And just in time. It will take all of our skills of communication to work through the fire of our collective initiation as we awaken to ourselves as a human community.

Internet use, while exploding, is generating a largely fragmented communication. Billions of individual messages lack effective ways of achieving a larger consensus, coherence, and visibility. Big-picture coherence currently comes from television. While it is tempting to regard television as an outmoded technology, the reality is that, despite Internet use going up, TV viewing has not gone down. In nearly every place on Earth, most people continue to get most of their news about the world from television. Because we require a new level of maturity as citizens and consumers, we also require a new level of responsibility in how we use television in synergy with the Internet. Three major changes are essential, immediately.

First, we require ecologically oriented advertising to balance the onslaught of pro-consumerist messages and to foster a mindset of authentic choice in our consumption behavior. To balance the psychological impact of the one-sided avalanche of commercials, we need "ecological ads" or "Earth-commercials" that encourage people to consume with an appreciation of their impact on the world's dwindling resources and deteriorating environment. Perhaps called "Earthvisions," they could be sixty-second ministories portraying some aspect of a sustainable and meaningful future and could be produced by individuals, nonprofit organizations, and local community groups working in partnership with local television stations. These ads for the Earth could be low in cost and high in creativity, and done with playfulness, compassion, and humor. Once under way, a virtual avalanche of "Earthvisions" could emerge from around the world and be shared over TV and the Internet.

Second, because television teaches continuously about the lifestyles and values that are the "norm" for society, we require entertainment programming that actively explores issues of sustainability and Earth-friendly approaches to living. Television teaches by what it ignores as well as by what it addresses. If an ecological consciousness and an ethic of sustainability are missing from entertainment programming, then they are likely to be missing from our cultural consciousness as well. We could get a tremendous cultural boost to sustainability from entertainment programming that explores ecological concerns and innovative ways of living.

Third, a mature citizenry requires expanded documentaries and investigative reports describing, in depth, the global challenges we now face. Because the overwhelming majority of prime-time hours on television are devoted to programming for amusement, we are entertainment-rich and knowledge-poor. Our situation is like that of a long-distance runner who prepares for a marathon by eating primarily junk food. We are filling our social brain with a diet of entertainment at the very time our democracies face problems of marathon proportions. We are trivializing our species consciousness at the very time we require mature communication about our pathway into the future. This is a recipe for disaster. We need a rapid and quantum increase in the level of ecologically relevant programming and a new social commitment to investigative journalism that awakens mainstream public attention to themes of sustainability.

Power of Democracy   Communication is the lifeblood of democracy. To choose a sustainable future, citizens must first be able to communicate among themselves about the future they want to bring into existence. We require a conscious democracy that pays attention to what is going on and that uses the modern tools of mass communication to enable citizens to engage in unprecedented levels of dialogue and consensus-building about our future. A healthy democracy requires the active consent of the governed, not simply their passive acquiescence. In some way, a maturing society must do more than use television as a one-way, passive medium for entertainment; instead, with the aid of the Internet, we can create two-way or interactive opportunities for citizen learning and dialogue—in short, "Electronic Town Meetings."

Democracy has often been called the art of the possible. If we don't know how our fellow citizens think and feel about policies to create a sustainable future, then we are floating powerless in a sea of ambiguity and unable to mobilize ourselves into constructive action. The most powerful and direct way to revitalize democracy is by improving the ability of citizens to know their own minds as an overall community, region, or nation. Trust in the good judgment of citizens seems well founded. For example, after a half century of polling U.S. public opinion, George Gallup concluded that the collective judgment of citizens was "extraordinarily sound." He also said that citizens are often ahead of their elected leaders in accepting innovations and radical changes.

Given that we can trust the collective wisdom of the body politic, we can cultivate that wisdom by engaging in regular Electronic Town Meetings at every scale—local, national, and global. By combining televised dialogues on key issues with instantaneous feedback from a scientific sample of citizens, the public can know its collective sentiments with a high degree of accuracy. Just as a doctor can take a very small sample of blood and obtain an accurate picture of the overall condition of one's body, so, too, can a relatively small, random sample of citizens be used to get a reliable picture of the strength, texture, and intensity of the views of the overall body politic.

With regular Electronic Town Meetings, or ETMs, combined with other types of forums at every scale, the perspectives and priorities of the citizenry could be brought into public view rapidly. As a working consensus emerges, it would presumably guide but not compel decision makers. The value of ETMs is *not* for citizens to micromanage government through direct democracy; rather, it is for citizens to discover the broadly shared values and priorities that can guide their representatives in government. Involving citizens in choosing the pathway into the future will not guarantee that the "right" choices will always be made, but it will guarantee that citizens feel involved and invested in those choices. Rather than feeling cynical and powerless, we will feel engaged and responsible for our future.

One of the biggest challenges facing humanity is to evolve the art and practice of conscious democracy in the communications era. To realize this challenge, we require a citizen-based "communication rights movement" that seeks a fair and just use of our tools of mass communication in service of a sustainable future for humanity. Each generation must renew its commitment to democracy in ways that respect the unique demands of the times. In this generation, our contract with democracy requires that citizens confront the unprecedented challenge of developing the tools and skills for mass social dialogue and consensus-building so we can democratically build a sustainable future.[52]

We cannot create a new world in the cultural context of old media programming. The old media is selling a culture of consumption. The new media must serve a culture of conservation and sustainability. As we think, so we will become. If we fill our social mind with old media, there is no room to imagine new possibilities. The wonderful thing about media is that it can change in the blink of an eye. For example, if the public wanted it, television broadcasters could open the airwaves to "Earthvisions" created by the youth of the world—appealing for a sustainable pathway for all. This could rapidly bring a new culture and consciousness into our lives and help shift our society in a new direction.

Power of Love   Currently, the human family is profoundly divided. We are divided between rich and poor; different racial, ethnic, and religious groups; men and women; current and future generations; humans and other species; geographic regions; and more. If we are to work together to build a sustainable future, we must learn to bridge and heal these differences. We will never create a sane and dynamically stable global community without healing these many divisions.

To achieve authentic and lasting reconciliation as the foundation for our future, we require compassion as a practical basis for organizing human affairs. Compassion is a realistic foundation for human relations, as it is a part of the "common sense" of humanity. Although this common sense may be variously stated, it is universally recognized:

Christianity

> As you wish that men would do to you, do so to them.
>
> —*Luke 6:31*

Buddhism

> Hurt not others in ways that you yourself would find hurtful.
>
> —*Udanavarga*

Judaism

> That which is hateful unto you, do not impose on others.
>
> —*Talmud, Shabbat 31a*

Hinduism

> Do naught unto others which would cause you pain if done to you.
>
> —*Mahabharata 5:1517*

Islam

> No one of you is a believer until he desires for his brother that which he desires for himself.
>
> —*Sunan*

Confucianism

> Do not unto others what you would not have them do unto you.
>
> —*Analects of Confucius 15:23*

These affirmations of a "golden rule" are like different facets of a single jewel, or like different branches of a single tree. At some level the human family already recognizes the immense practical value of compassion as a basis for human affairs.

We all have our unique work to do and our unique contributions to make to life. If each person were to begin to act in a more compassionate manner, no matter how imperfect or tentative his or her actions, the cumulative result for the entire society would be enormous. Bit by bit, small changes could accumulate into a tidal wave of revitalizing change. A practical expression of the "golden rule" could be to confront the inequities in the world and to make good-faith efforts to consume in ways that might be generalized to all people in the world.

We may imagine love to be quite utopian, but consider the alternatives. In not choosing love we are left with law and the prospect of global bureaucratic stagnation. In not choosing law we are left with force and the prospect of either global devastation or global domination. If we value our freedom and vitality as a species, we are obliged to do no less than learn to love one another as a human family. With the prospect of genuine reconciliation, we can begin a historic process of healing that will enable us to honor our differences *and* work together for a future that benefits us all. Global reconciliation and cooperation offer a practical and promising pathway into the future.

To live *sustainably*, we must live efficiently—not misdirecting or squandering the Earth's precious resources. To live *efficiently*, we must live peacefully, for military expenditures represent an enormous diversion of resources from basic human needs. To live *peacefully*, we must live with a reasonable degree of *equity*, or fairness, for it is unrealistic to think that, in a communications-rich world, several billion persons will accept living in absolute poverty while another billion lives in extravagant excess. Only with greater fairness in the consumption of the world's resources can we live peacefully, and thereby sustainably, as a human family. Without a revolution in fairness based upon an awakening of social compassion, the world will find itself in profound conflict over dwindling resources like arable land and freshwater. A world in conflict seems unlikely to be able to mobilize itself quickly and respond to such critical problems as climate change and the end of cheap oil. Therefore, only with greater equity can we expect to live peacefully, and only with greater compassion can we expect to live sustainably.

If the world is profoundly divided materially, there is very little hope that it can be united socially, psychologically, and spiritually. Therefore, if we intend to live together peacefully as members of a single, human family, then each individual needs to expect a fair share of the world's wealth, sufficient to support a "decent" standard of living—one that provides enough food, shelter, education, and health care to enable everyone to realize their potential as productive and respected members of the human community. This does not mean that the world should adopt a single manner and standard of living; rather, it means that each person needs to feel part of the global family and, within a reasonable range of differences, valued and supported in realizing his or her unique human potential.

The "cost of compassion" is surprisingly low: The United Nations "Human Development Report of 1998" reported that expenditures for pet food, perfume, and ice cream in developed nations vastly exceeded the total resources needed to eliminate world hunger, immunize every child, provide clean drinking water and sanitation for all, and offer universal education. If we live moderately, we have the material means to establish a decent standard of living for everyone.

In considering these six great empowerments and how powerfully they work together as an integrated system, it becomes clear they have the potential to transform adversity into unprecedented opportunity. We tend to think that we are powerless, helpless, impotent. Yet the reality is that only we—as individuals working in cooperation with one another—have the power to transform our situation. Far from being helpless, we are the only source from which the necessary creativity, compassion, and action can arise. The outcome of this time of planetary transition will depend on the choices that we make as individuals. Nothing is lacking. Nothing more is needed than what we already have. We require no remarkable, undiscovered technologies. We do not need heroic, larger-than-life leadership. The only requirement is that we, as individuals, choose a revitalizing future and then work in community with others to bring it to fruition. By our conscious choices we can move from alienation to community, from despair to creativity, from passivity to participation, from stagnation to learning, from cynicism to caring.

THE PROMISE OF SIMPLICITY

In closing, I want to return to a core theme from the beginning of this book: When we reflect on the powerful forces transforming our world—climate change, peak oil, water shortages, species extinction—it is clear that we require far more than cosmetic changes in our manner of living. We require a conscious and deep simplicity if we are to rise to meet the challenges of the gathering world storm and build a workable future. Conscious simplicity is not an alternative way of life for a marginal few; it is a creative choice for the mainstream majority, particularly in developed nations. If we are to pull together as a human community, the path of a deep and elegant simplicity is vital for establishing a secure foundation for sustainability. Simplicity is not only a personal choice—it is also a choice for nations and the entire species.

The human family has entered a transitional time. An enormous distance has been traveled over the past several hundred years—socially, psychologically, technically, and politically. We are entering an entirely new situation in human affairs as we encounter ourselves at a global level. As a species, we are like adolescents who face our early adulthood and the necessity of assuming the difficult responsibilities of a new level of collective maturity. As we look ahead, we see a future that is radically uncertain. The entire world is in transition, trying to makes its way into a future that is still barely comprehensible. Given the confusion and chaos of our pivotal times, we may feel drawn to the certainties of the past. But it is now our responsibility to collaborate as a human community in meeting the challenge of building a sustainable and meaningful future. Simplicity of living both enables this journey and is an expression of its fulfillment. Whole new dimensions of human opportunity await us if we will rise to the challenge of living more consciously and simply.

The philosopher Søren Kierkegaard said, "Hope is passion for the possible." When I look at the challenges facing the human family, I ask myself: Is it possible to stabilize the global climate? With heroic efforts and sustained cooperation, yes we can. Is it possible for the world to make a swift transition to renewable sources of energy? With a daring level of global collaboration, yes we can. Is it possible to restore vital habitats on the land and in the oceans so that other species may flourish? With teamwork, yes we can. Is it possible to enter into global conversation and discover a new story that honors our past and guides us into a promising future? With respect for the magnificence of the human journey, yes we can. Overall, the perfect storm of a world systems crisis presents humanity with a perfect opportunity for making a collective leap forward. I look at the future with a realistic sense of hope. I know that it is possible for the human family to say, "Yes we can!" and, with conscious simplicity, cooperation, and bold effort, build a sustainable and promising future.

Simplicity Resources

BOOKS *There is a rapidly growing literature on simplicity and sustainable living. Below is a diverse collection of books that I've found particularly valuable.*

Cecile Andrews, *The Circle of Simplicity: Return to the Good Life*, New York: HarperCollins, 1997. This book has been a catalyst for study circles around the world. A wise and warm exploration of the path to a simpler life and a helpful guide to conversations.

Mark Burch, *Simplicity: Notes, Stories and Exercises for Developing Unimaginable Wealth*, Philadelphia: New Society Publishers, 1995. An eloquent and soulful guide for reflecting on our lifestyle, with practical exercises for personal and community-wide change.

Joe Dominguez and Vicki Robin, *Your Money or Your Life*, New York: Viking Press, 1992, revised 2008. A cornucopia of insight about transforming your relationship with money and achieving financial independence. Energized with the personal stories of diverse people, this book describes an inspiring path for living with frugality, integrity, and compassion.

Michael Gellert, *The Way of the Small: Why Less Is Truly More*, Florida: Nicolas-Hays, 2008. A practical and spiritual guide to fulfillment; Gellert reframes the search for happiness, meaning, and success and writes that happiness is to be found in "the small"—the precious gifts of ordinary life.

Paul Hawken, *Blessed Unrest: How the Largest Movement in the World Came into Being and Why No One Saw It Coming*, New York: Viking Press, 2007. Describes the leaderless revolution occurring around the world involving activists for environmental sustainability, social justice, and indigenous values.

Ellis Jones, Ross Haenfler, and Brett Johnson, *The Better World Handbook: From Good Intentions to Everyday Actions*, Canada: New Society Publishers, 2007. A definitive guide for the average person wanting to make a positive difference in the world. Specifically designed for well-intentioned people who may be too busy to be actively involved in social change organizations.

David Korten, *The Great Turning: From Empire to Earth Community*, San Francisco: Berrett-Koehler, 2006. This masterful integration of economics, psychology, spirituality, and more presents a blueprint for a sustainable and meaningful future.

Joel and Michelle Levey, *Living in Balance: A Dynamic Approach for Creating Harmony and Wholeness in a Chaotic World*, Berkeley, California: Conari Press, 1998. A balanced life equals health—mental, emotional, and physical. Two experts offer a synthesis of ancient meditative traditions with cutting-edge research on peak human performance to show readers how to master the art of balance within an environment of rapid change.

Janet Luhrs, *The Simple Living Guide: A Sourcebook for Less Stressful, More Joyful Living*, New York: Broadway Books, 1997. With wit and wisdom, the author presents stories, strategies, and resources for home, family, work, holidays, and more to create an empowering guide for anyone seeking a simpler life.

Linda Breen Pierce, *Choosing Simplicity: Real People Finding Peace and Fulfillment in a Complex World*, Carmel, California: Gallagher Press, 2000. Linda brings clarity and compassion to an insider's view of the simple life based upon a three-year study of several hundred people who simplified their lives.

—. *Simplicity Lessons: A 12-Step Guide to Living Simply*, Carmel, California: Gallagher Press, 2003. A practical guide for simplifying one's life and having more time for relationships, fulfilling work, and satisfying living. An important resource for both individuals and study groups.

David Shi, *The Simple Life: Plain Living and High Thinking in American Culture*, New York: Oxford University Press, 1985. An in-depth exploration of the history of the simple life in the United States—from the Puritans and Quakers to the Transcendentalists and Progressives, and beyond.

James Gustave Speth, *The Bridge at the Edge of the World: Capitalism, the Environment, and Crossing from Crisis to Sustainability*, New Haven: Yale University Press, 2008. A practical and wise view on the failures of capitalism and the steps needed to transform the economy and the political process for building a sustainable future.

Alex Steffen, ed., *World Changing: A User's Guide for the 21st Century*, New York: Abrams, 2006. Filled with ideas, solutions, and stories that offer innovative solutions for tackling the challenges of sustainability.

Liz Walker, *EcoVillage at Ithaca: Pioneering a Sustainable Culture*, Canada: New Society Publishers, 2005. An internationally recognized model of sustainable development, with innovations including cohousing neighborhoods, small-scale organic farming, green energy, and hands-on education. An inside look at new community pioneers for a sustainable future.

Roger Walsh, *Essential Spirituality: The 7 Central Practices to Awaken Heart and Mind*, New York: John Wiley & Sons, 1999. Based on more than twenty years of research, Walsh explores seven inner practices found in all of the world's great spiritual traditions. This wise and helpful book is filled with exercises, meditations, stories, prayers, and practical insights.

David Wann, ed., *Reinventing Community: Stories from the Walkways of Cohousing*, Colorado: Fulcrum Publishing, 2005. Cohousing is a way to re-create community in the modern world. Wann has brought together stories from the real-world perspectives of people who live in these communities, whether urban, suburban, or rural.

—. *Simple Prosperity: Finding Real Wealth in a Sustainable Lifestyle*, New York: St. Martin's Press, 2007. Explores how we can have a more abundant and sustainable lifestyle without sacrificing everything we love.

WEBSITES *There are a huge number of websites dealing with various facets of sustainability and green living. Because this area is growing so rapidly, below I list a few key websites that can serve as portals to many other Internet resources:*

The Simple Living Network: www.simpleliving.net
   The Simple Living Network is the premier online simple living resource—a small, home-based cottage business operated by Dave Wampler in the Cascade Mountain village of Trout Lake, Washington. Since 1996, this network has been providing resources, tools, examples, and contacts for conscious, simple, healthy, and restorative living. Through it, you can gain access to many other learning materials, including resources specifically tailored for use in group study situations—for example, the "circles of simplicity" developed by Cecile Andrews; the study guide to accompany the book *Your*

Money or Your Life, coauthored by Joe Dominguez and Vicki Robin; and much more.

Awakening Earth: www.awakeningearth.org

This is my personal website and offers books, articles, videos, and links in four major areas: 1) the big picture of the human journey with the world at the tipping point; 2) the importance of simplicity, sustainability, and community for building a green future; 3) the universe as a living system and how we can live in conscious relationship with it; and 4) the role of media accountability and citizen empowerment in responding to the challenges and opportunities of our times.

The Northwest Earth Institute: www.nwei.org

NWEI develops innovative programs that empower individuals and organizations to move toward a sustainable and enriching future. It also offers study guides for small groups. These self-guided discussion courses are given in workplaces, universities, homes, faith centers, neighborhoods, and community centers throughout North America. Each program encourages participants to explore values, attitudes, and actions through discussion with other people.

WiserEarth: www.wiserearth.org

WiserEarth is a free-to-use, noncommercial, online community space that maps and connects organizations and individuals who are addressing the central issues of our day: climate change, poverty, the environment, peace, water, hunger, social justice, conservation, human rights, and more. Here you will find a huge directory of nongovernmental organizations (NGOs) and socially responsible organizations, as well as online community forums where members can engage in discussion, post and share resources, and collaborate on projects.

Global Ecovillage Network: gen.ecovillage.org

The Global Ecovillage Network is a global confederation of people and communities that meet and share their ideas; swap technology information; develop cultural and educational exchanges, directories, and newsletters; and are dedicated to restoring the land and living "sustainable plus" lives by putting more back into the environment than we take out.

Transition Towns: www.transitiontowns.org

This site is a Wiki for use by all the communities that have adopted the "Transition Model" for responding positively and creatively to the twin challenges of peak oil and climate change. It provides a focal point for all of these towns, villages, cities, and localities around the world as they implement their own "Transition Initiative." Some communities are using this Wiki as their main website for organizing, recording progress, and communicating with their own community. Others have developed their own websites.

Peter Russell: www.peterrussell.com/index2.php

This wide-ranging website explores the transition to a more conscious culture and sustainable future. In addition to his own writing on consciousness, science, and transformation, Peter includes extensive links to other resources concerned with an awakening consciousness and culture.

Notes

1. Jared Diamond, *Collapse*, New York: Viking Press, 2005, p. 498.

2. The "Warning to Humanity" was sponsored by the Union of Concerned Scientists, 26 Church St., Cambridge, Massachusetts 02238.

3. Paul Hawken, *Blessed Unrest*, New York: Viking Press, 2007. Summary taken from the press materials that accompanied his book.

4. Quoted in: David Shi, *The Simple Life: Plain Living and High Thinking in American Culture*, New York: Oxford University Press, 1985, p. 145.

5. Frank Lloyd Wright quoted in Shi, ibid., p. 187.

6. The following surveys illustrate how simplicity and sustainability have become an integral part of a mainstream culture, involving millions of people. Gerald Celente, president of the Trends Research Institute, reported in 1997 on how the voluntary simplicity trend is growing throughout the industrialized world: "Never before in the Institute's 17 years of tracking has a societal trend grown so quickly, spread so broadly and been embraced so eagerly" (Gerald Celente, *Trends Journal*, Winter 1997). The following surveys provide further evidence that a lifeway of conscious simplicity, with a characteristic pattern of values, is emerging as a significant trend in the world.

Yearning for Balance: A 1995 survey of Americans commissioned by the Merck Family Fund found that respondents' deepest aspirations are nonmaterial (*Yearning for Balance: Views of Americans on Consumption, Materialism, and the Environment*, a report by the Harwood Group about a survey conducted for the Merck Family Fund, Takoma Park, Maryland, July 1995). For example, when asked what would make them much more satisfied with their lives, 66 percent said "if I were able to spend more time with my family and friends," and only 19 percent said "if I had a bigger house or apartment." Twenty-eight percent of the survey respondents said that over the last five years they had voluntarily made changes in their lives that resulted in their earning less money, such as reducing work hours, changing to a lower-paying job, or even quitting work. The most frequent reasons given for voluntarily downshifting were: 1) wanting a more balanced life (68 percent), 2) wanting more time (66 percent), and 3) wanting a less stressful life (63 percent). Had it been worth it? Eighty-seven percent of the downshifters described themselves as happy with the change. In summing up the survey's findings, the report stated, "People express a strong desire for a greater sense of balance in their lives—not to repudiate material gain, but to bring it more into proportion with the non-material rewards of life."

World Values Survey: This massive survey concluded that over the last three decades a major shift in values has been occurring in a cluster of a dozen or so nations, primarily in the United States, Canada, and northern Europe. See: Inglehart, Foa, Peterson, and Welzel, "Development, Freedom, and Rising Happiness: A Global Perspective (1981–2007)," *Association for Psychological Science*, Vol. 3, No. 4, 2008. Ronald Inglehart, "Changing Values Among Western Publics from 1970 to 2006," *West European Politics*, January–March 2008.

Health of the Planet Survey: In 1993, the Gallup organization conducted in twenty-four nations this landmark global survey of attitudes toward the environment (see: Riley E. Dunlap, "International Attitudes Towards Environment and Development," in *Green Globe Yearbook 1994*, an independent publication from the Fritjof Nansen Institute, Norway, Oxford University Press, 1994, p. 125). In writing about the survey, institute director Dr. Riley E. Dunlap concluded that residents of poorer and wealthier nations express nearly equal concern about the health of the planet. Majorities in most of the nations surveyed gave environmental protection a higher priority than economic growth, and said that they were willing to pay higher prices for that protection. There was little evidence of the poor blaming the rich for environmental problems, or vice versa. Instead, there seemed to be a mature and widespread acceptance of mutual responsibility. When asked who was "more responsible for today's environmental problems in the world," the most frequent response was that industrialized and developing countries were "both equally responsible."

World Environmental Law Survey: The largest environmental survey ever conducted was done in the spring of 1998 for the International Environmental Monitor. Involving more than thirty-five thousand respondents in thirty countries, the survey found that "majorities of people in the world's most populous countries want sharper teeth put into laws to protect the environment." Majorities in twenty-eight of the thirty countries surveyed (ranging from 91 percent in Greece to 54 percent in India) said that environmental laws as currently applied in their country "don't go far enough." The survey report concluded, "Overall, these findings will serve as a wake-up call to national governments and private corporations to get moving on environmental issues or get bitten by their citizens and consumers, who will not stand for inaction on what they see as key survival issues."

7. Ibid.

8. Duane Elgin and Arnold Mitchell, "Voluntary Simplicity," Business Intelligence Program, Long-Range Planning Service, SRI, Menlo Park, California, June 1976, No. 1004.

9. Michael S. Rosenwald, "Showcasing the Growth of the Green Economy," *Washington Post*, October 16, 2006, p. D01.

10. Linda Breen Pierce, *Choosing Simplicity*, Carmel, California: Gallagher Press, 2000.

11. www.worldvaluessurvey.com.

12. Ronald Inglehart, op. cit.

13. I am grateful to Arnold Mitchell for suggesting this example to me.

14. Roger Walsh, "Initial Meditative Experiences: Part I," *Journal of Transpersonal Psychology*, No. 2, 1977, p. 154.

15. See, for example: Chogyam Trungpa, "Foundations of Mindfulness," in *Garuda IV* (Berkeley, California: Shambala Press, 1976); Sri Nisargadatta Maharaj, *I Am That*, Vols. I and II, Maurice Frydman (trans.) (Bombay, India: Chetana, 1973); Joseph Goldstein, *The Experience of Insight* (Santa Cruz, California: Unity Press, 1976).

16. See, for example: Aldous Huxley, *The Perennial Philosophy* (New York: Harper, 1945); Huston Smith, *The Religions of Mankind* (New York: Harper & Row, 1958).

17. Richard Gregg, "Voluntary Simplicity," reprinted in *Co-Evolution Quarterly*, Sausalito, California, Summer 1977; originally published in the Indian journal *Visva-Bharati Quarterly*, August 1936.

18. See, for example: Inglehart, Foa, Peterson, and Welzel, "Development, Freedom, and Rising Happiness: A Global Perspective (1981–2007)," *Association for Psychological Science*, Vol. 3, No. 4, 2008. Ronald Inglehart, "Changing Values Among Western Publics from 1970 to 2006," *West European Politics*, January–March 2008. Ed Diener and Martin Seligman, "Beyond Money: Toward an Economy of Well-Being," *American Psychological Society*, Vol. 5, No. 1, 2004. Tim Kasser, *The High Price of Materialism*, Cambridge, Massachu-

setts: MIT Press, 2002. Bill McKibben, "Reversal of Fortune," *Mother Jones*, March-April 2007.

19. Kasser, op.cit., p. 14.

20. Ibid., p. 22.

21. James Fowler and Nicholas Christakis, "Dynamic spread of happiness in a large social network: longitudinal analysis over 20 years in the Framingham Heart Study," *BMJ* 2008, 337:a2338.

22. James Fowler, quoted in: Rob Stein, "Happiness Can Spread Among People Like Contagion, Study Indicates," www.washingtonpost.com, December 5, 2008.

23. Arnold Toynbee, *A Study of History* (Abridgement of Vols. I–VI, by D.C. Somerville), New York: Oxford University Press, 1947, p. 198.

24. These questions were taken from an early version of a book written by the Simple Living Collective of San Francisco: *Taking Charge*, New York: Bantam Books, 1977.

25. Ramana Maharshi on silence.

26. Gregg, op. cit., p. 20.

27. Thoreau quoted in Shi, op. cit., p. 149.

28. Gandhi quoted in Gregg, op. cit., p. 27.

29. See: WWF report, "Climate Change: Faster, stronger, sooner," a European update of climate science, 2008.

30. See the 350 initiative launched by Bill McKibben: www.350.org.

31. There is an enormous amount of literature on the end of cheap oil but relatively little exploration of what life will be like beyond the petroleum age. Here are three books I recommend: James Speth, *The Bridge at the Edge of the World*, New Haven: Yale University

Press, 2008. David Korten, *The Great Turning*, San Francisco: Berrett-Koehler, 2006. James Kunstler, *The Long Emergency*, New York: Grove Press, 2006.

32. Sandra Postel, "Water for Food Production: Will There Be Enough in 2025?" *BioScience*, Vol. 48, No. 8, August 1998.

33. See, for example: Joby Warrick, "A Warning of Mass Extinction," *Washington Post*, April 21, 1998; Virginia Morell, "The Sixth Extinction," *National Geographic*, February 1999.

34. See: WWF, "Living Planet Report, 2008," which looks at humanity's ecological footprint. Available online at: www.panda.org/news_facts/publications/living_planet_report/lpr_2008.

35. "Special report: How our economy is killing the Earth," *New Scientist*, October 16, 2008.

36. David Shi's book *The Simple Life* was invaluable in developing this historical overview.

37. Quoted in Goldian VandenBroek, ed., *Less Is More*, New York: Harper Colophon Books, 1978, p. 116.

38. Ibid., p. 60.

39. Shi, op. cit., p. 4.

40. Ibid., p. 127.

41. Duane Elgin, *The Living Universe*, San Francisco: Berrett-Koehler, 2009.

42. David Bohm, *Wholeness and the Implicate Order*, London: Routledge & Kegan Paul, 1980, p. 11.

43. Ibid., p. 191.

44. Brian Swimme, *The Hidden Heart of the Cosmos*, New York: Orbis Books, 1996, p. 100.

45. See the discussion in my book *Promise Ahead*, New York: Quill Books, 2000, pp. 52–57.

46. Matthew Fox, *Meditations with Meister Eckhart*, Santa Fe: Bear & Co., 1983, p. 24.

47. Smith, *The Religions of Man*, op. cit., p. 73.

48. D. T. Suzuki, *Zen and Japanese Culture*, New York: Pantheon Books, 1959, p. 364.

49. Lao-tzu, *Tao Te Ching* (translation by Gia-Fu Feng and Jane English), New York: Vintage Books, 1972.

50. See: "Americans Can't Get Enough of Their Screen Time." This report on TV, Internet, and mobile use found that TV viewing was at an all-time high despite continued increases in the use of the Internet and mobile devices. News release: The Nielsen Company, New York, November 24, 2008. Also see: "Key News Audiences Now Blend Online and Traditional Sources," Pew Research Center Publications, August 17, 2008.

51. Transition Handbook; see: transitionculture.org/shop/the-transition-handbook.

52. When the public feels enough pressure and pain, we will reclaim the rights of citizenship and take back the airwaves from advertisers and begin to use them for sustained and searching dialogue about our pathway into the future—locally, nationally, and globally. In the United States, *the airwaves are owned by the public.* Although the public does not realize this and feels victimized by broadcasters, the legal reality is that the public is the owner of the airwaves and broadcasters are obligated "to serve the public interest, convenience, and necessity" before they serve their corporate profits. Because the public is ignorant of its ownership and its authority regarding how the airwaves are used, citizens are needlessly disempowered. The public may blame the national networks for programming, but it is the met-

ropolitan-scale TV broadcasters that use the public's airwaves and are legally obligated to serve the interests of their local community. A "community voice" organization whose scale of trans-partisan representation matches the scale of the media footprint of television broadcasters can have an immediate and profound impact by, for example, hosting regular electronic town meetings concerned with how the public's airwaves are being used. For example, does the community want to see a different mix of programming that gives significantly more attention to the climate crisis, the longer-range future, and green ways of living?

Index